THEORETICAL PHYSICS AND PHILOSOPHICAL PROBLEMS

VIENNA CIRCLE COLLECTION

VOLUME 5

LUDWIG BOLTZMANN (1844–1906)

LUDWIG BOLTZMANN

THEORETICAL PHYSICS
AND
PHILOSOPHICAL PROBLEMS

Selected Writings

Edited by

BRIAN McGUINNESS

With a Foreword by

S. R. DE GROOT

D. REIDEL PUBLISHING COMPANY

DORDRECHT-HOLLAND / BOSTON-U.S.A.

Translations from the German by Paul Foulkes

Library of Congress Catalog Card Number 74-79571

Cloth edition: ISBN 90 277 0249 7
Paperback edition: ISBN 90 277 0250 0

Published by D. Reidel Publishing Company,
P.O. Box 17, Dordrecht, Holland

Published in the U.S.A., Canada, and Mexico
by D. Reidel Publishing Company, Inc.
306 Dartmouth Street, Boston,
Mass. 02116, U.S.A.

Printed in The Netherlands by D. Reidel, Dordrecht

CONTENTS

PART I / FROM *POPULÄRE SCHRIFTEN*
(Writings addressed to the Public)

PART II / FROM *NATURE* **51** (1895)

FOREWORD

1. The work of Ludwig Boltzmann (1844–1906) consists of two kinds of writings: in the first part of his active life he devoted himself entirely to problems of physics, while in the second part he tried to find a philosophical background for his activities in and around the natural sciences.[1] Most scientists are much more aware of his creative work in physics than of his digressions on the meaning and structure of science. I think in the present case the reason is not so much that most scientists are usually almost entirely occupied with their trade, because Boltzmann's philosophical work is also concerned with the (natural) sciences. I rather believe that the quality and consistency of Boltzmann's purely scientific work is of a more appealing nature than his less structured considerations on human activity in science and in life in general.

2. I think that it may be appropriate for the readers of this anthology to say a few words on the main findings of Boltzmann in physics, since in the end their 'philosophical' impact has been larger than the effect of his later writings. Moreover some knowledge of his scientific achievements can be helpful for the understanding and appreciation of the essays printed in this book, which almost all stem from Boltzmann's philosophical period.

Boltzmann was one of the main protagonists – at least in continental Europe – of atomistics for explaining the phenomena of physics. His fame, until this day, is mostly based on two theories. The first is his interpretation of the notion of 'entropy' as a mathematically well-defined measure for what one might call the 'disorder' or 'probability' of a collection of atoms. His ideas on this topic gradually evolved from tentative ones of a purely mechanical character to the final concept of a statistical property. In physical processes the systems moved from states to which he assigned a certain probability to states of ever greater probability. This property, according to which, in physical processes taking place in isolated systems, the entropy never decreases, is the celebrated 'H-theorem' of Boltzmann. It explains the irreversibility of the phenomena concerned as an essentially statistical property of a set of many atoms. In this way he solved one of the main problems of physics: how could systems of which the individual particles obeyed mechanical laws that were invariant under a change of the direction of time, show a behaviour with a definite preference for a development in time in one particular direction? The rebuttals, which did not fail to be raised, were answered by Boltzmann in sometimes rather violent discussions. It is true that Boltzmann, when refining his arguments, sometimes changed his views without stating so explicitly. Another feature of Boltzmann's style which rebuffed scientists belonging to different traditions was the often excessive length of his argumentations. James Clerk Maxwell, whom Boltzmann greatly admired, especially for the mechanical models which the famous Scot used in his theories, and who shared Boltzmann's atomistic views, once wrote to a friend and fellow-physicist: "By the study of Boltzmann I have been unable to understand him. He could not understand me on account of my shortness, and his length was and is an equal stumbling block for me. Hence I am very much inclined to join the glorious company of the supplanters and to put the whole business in about six lines." In fact Maxwell's writings are often succinct although his 'A Treatise on Electricity and Magnetism' proved to be difficult reading for not a few physicists for rather the opposite reason....

The second principal claim to lasting fame is the celebrated 'transport equation', which Boltzmann wrote down in 1872. It is frequently called the 'Boltzmann equation' and it provides a means to treat a number of properties of a gas, i.e., an example of a system consisting of a great number of particles. The model described by the Boltzmann equation represents a simplified description of a gas which on the one hand remains sufficiently near to a complete description to yield physically relevant results and which on the other hand permits a mathematical treatment that leads to predictions of physical properties. The equation contains 'distribution functions' which give the statistical probability of finding the atoms at certain positions in space-time and animated by certain velocities. It also contains a function which describes the collisions of the atoms and which is therefore characteristic for the particular system studied. Thus Boltzmann's equation involves both his main notions of physical theory: atomism and statistics. The equation allows the calculation of the so-called 'transport properties' of gases, such as their heat conductivities, viscosities and the like. It remains, after a century, an essential tool of physics. Its applications have been numerous and many new ones are added every year.

3. Boltzmann's academic career[3, 4, 5] reflects his preoccupations with physics, both theoretical and experimental, mathematics, and philosophy. In 1869, at the age of twenty-five, he became professor of mathematical physics in the university of Graz. In 1873 he moved back to his native Vienna to teach mathematics. In 1876 he went again to Graz, this time as a professor of experimental physics. In 1890 he accepted a professorship of theoretical physics in Munich. He returned again to Vienna in 1894, to leave for Leipzig in 1900, until finally in 1902 he settled definitely in Vienna. He succeeded Ernst Mach, who taught in Vienna from 1895 to 1901, and who had been his fierce opponent by attacking atomistics. The alternation between Austrian and German universities gave him some difficulties. Thus for instance the Austrian authorities reproached him for his absence from their country and made him feel this by a tardy renomination as a member of the Imperial Academy and by budgetary measures which caused practical difficulties and delays in scientific plans.

4. The second half of Boltzmann's active life was centred on philosophy, rather than on physics, albeit that the foundations of physical theory, in particular the doctrine of atomism, played an outstanding role in his papers, courses and conferences. Roughly speaking one can say that his work on physics lasted until about 1898, while his philosophical interest started to be important from 1886 on. The present book contains mainly writings of the philosophical period. For Boltzmann a theory was an analogy or a metaphor for reality. The mechanical models which he used did not need to be taken as 'reality'; he did not assert that a gas consisted 'really' of atoms and that the atoms 'really' interact. I agree in this view with the historian Martin J. Klein[4, 7]. I surmise that the editors of this collection are of the same opinion, since otherwise it would be hard to include Boltzmann in the Vienna Circle. However, other authors, especially Boltzmann's biographer, E. Broda, consider Boltzmann as a realist or a materialist who believed in the objective existence of an external world[2, 6]. Broda puts Boltzmann in contrast to Mach, for whom theories were merely orderings of sensations. Perhaps the difference between the two

points of view just mentioned is smaller than it seems from the tags employed, because, as Broda also stressed, Boltzmann claimed that the main merit of the theory of atomism was its capacity of explaining many phenomena, and of predicting numerous details which escaped the theories of the 'energeticists', who refused to employ the notion of atom. The brain, Boltzmann said, makes pictures of the world, which are useful representations of experiences. Theory is pre-eminently a practical matter: it possesses precision and allows us to check the validity of models and reasonings by numerical comparison with the experimental facts. Boltzmann was a strong adversary of the German idealists. He combated the views of Hegel, Schopenhauer and Kant. He said he found more philosophy in the concepts of the physicists than in those of the idealists.

Boltzmann had a tremendous admiration for Darwin and he wished to extend Darwinism from biological to cultural evolution. In fact he considered biological and cultural evolution as one and the same thing. The evolution of theories and ideas took place in successive jumps, just as biological evolution did. The pictures which evolved in the brain tended to perfection in the course of the centuries according to the same rules as laid down in Darwin's theory. Thus they developed slowly as representations of experiences. In short, cultural evolution was a physical process taking place in the brain. Boltzmann included ethics in the ideas which developed in this fashion and he considered his political views, those of a radical democrat and a republican, as outcomes of his ethical ideas.

5. At the end of Boltzmann's life the attacks on atomistics, at least in central Europe, grew powerful. It was only a decade or two later that atomic theory became preponderant. Today the relevance of Boltzmann's ideas in physics is clearer than ever and his fame is established for good. The same cannot be said of his influence in philosophical circles, either during his life, or thereafter.

Institute of Theoretical Physics,   S. R. DE GROOT
University of Amsterdam,
Amsterdam, The Netherlands

REFERENCES

[1] L. Boltzmann, *Wissenschaftliche Abhandlungen*. Three volumes. Editor: F. Hasenöhrl (J. A. Barth, Leipzig, 1909). Contains all of Boltzmann's work, except for the following five publications.

L. Boltzmann, *Vorlesungen über die Maxwellsche Theorie der Elektrizität und des Lichtes*. Two volumes (J. A. Barth, Leipzig, 1891, 1893).

L. Boltzmann, *Vorlesungen über Gastheorie*. Two volumes (J. A. Barth, Leipzig, 1895, 1898).

L. Boltzmann, *Vorlesungen über die Prinzipe der Mechanik*. Two volumes (J. A. Barth, Leipzig, 1897, 1904).

L. Boltzmann, *Populäre Schriften* (J. A. Barth, Leipzig, 1905).

L. Boltzmann and J. Nabl, 'Kinetische Theorie der Materie', *Enzyklopädie der mathematischen Wissenschaften V* 8, 1907 (B. G. Teubner, Leipzig-Vienna, 1970).

[2] E. Broda, *Ludwig Boltzmann, Mensch-Physiker-Philosoph* (F. Deuticke, Vienna, 1955).

[3] R. Dugas, *La théorie physique au sens de Boltzmann et ses prolongements modernes* (Le Griffon, Neuchâtel, 1959).

[4] Martin J. Klein, *Paul Ehrenfest*, Volume 1, *The Making of a Theoretical Physicist* (North-Holland Publishing Company, Amsterdam-London, 1970).

[5] D. Flamm, 'Life and Personality of Ludwig Boltzmann', in *The Boltzmann Equation*. Editors: E. G. D. Cohen and W. Thirring, *Acta Physica Austriaca Suppl.* 10 (1973) 3 (Springer Verlag, Vienna-New York, 1973).

[6] E. Broda, 'Philosophical Biography of L. Boltzmann', *ibid.*, p. 17.

[7] M. J. Klein, 'The Development of Boltzmann's Statistical Ideas', *ibid.*, p. 53.

EDITOR'S NOTE

The first aim of the following selection is to include all of Boltzmann's *Populäre Schriften* (his *Writings addressed to the Public*) that will convey to the modern reader his individual conception of the nature of science in general and theoretical physics in particular. This was the aspect of his thought which, alongside his own contributions to theoretical physics, attracted the attention of members of the Vienna Circle and of related thinkers such as Wittgenstein, and thus won for Boltzmann, as one of the least of his posthumous guerdons, a place in this series.

Accordingly it has been seen fit to omit all pieces expounding the ideas or the merits of others. Under this head fall: (2) 'On Maxwell's Theory of Electricity' (1873), (4) 'Gustav Robert Kirchhoff' (1887), (7) 'Josef Stefan' (1895)' (13) 'Röntgen's New Rays' (1896), (15) 'In Memory of Josef Loschmidt' (1895), and (21) 'Review of W. Vaubel's Textbook of Theoretical Chemistry' (1903). Many of these might have been included had not that economy in printer's ink which Frege once derided become a necessity. It may perhaps be mentioned that (15) contains Boltzmann's endorsement of Loschmidt's ironical suggestion of a negative scientific journal which would record only unsuccessful experiments. Loschmidt, he says, had ideas which, with slightly better experimental means, would have anticipated important discoveries, but the suggestion has wider applications. One polemical piece has been omitted because its technical character and its close involvement with contemporary publications make it chiefly valuable to specialists in the controversy. This is the 33-page essay, (8) 'A Word from Mathematics to Energetics' (1896). The last two lectures of (16) have likewise been omitted because of their technical character.

The remaining two omissions are (6) 'On Airship Flight' (1894), and (23) 'A German Professor's Journey to Eldorado' (1905). The former correctly predicts the superiority of the aeroplane over the airship. The latter records Boltzmann's visit to the Berkeley of staider days. It contains many flights of his sense of humour, but the attentive reader of the present selection will not feel cheated in that regard. The tone of Boltzmann's writings, it may be remarked, is strikingly at variance with his tragic end.

The remainder of the present selection is made up of two articles not included in *Populäre Schriften*, which nevertheless give Boltzmann's views on general issues in a succinct form, and of a selection with, it is hoped, similar properties from his *Vorlesungen über die Principe der Mechanik (Lectures on the Principles of Mechanics)*.

The two articles last mentioned were first published in English. All other translation in this volume is the work of Dr Paul Foulkes, except for that of the dedication and preface to *Populäre Schriften*, which is the present writer's. In some cases a correct German text had to be restored: wherever there was the slightest room for doubt, the nature of the correction has been indicated in a footnote. Dr Foulkes has made explicit in his translation some literary allusions which the original readers were expected to recognize without cues.

B. MCGUINNESS

PART I

From *Populäre Schriften*

WRITINGS ADDRESSED
TO THE PUBLIC

Dedicated
a century after
his entry into immortality
to the shade of Schiller
unsurpassed master of depiction faithful to nature
and of true enthusiasm arising from the depths of the heart

FOREWORD

[In an opening paragraph Boltzmann explains that, unable at his age to learn the spelling introduced by a recent reform, he has adopted for the foreword an entirely phonetic orthography. "aidhə let dhə dog keep itz teil aw kʌt dhə houl thing awf aet wʌn gou!" In translation the tail has been left alone.]

In the present volume I have, at the publisher's invitation, assembled my writings addressed to the public. They are of very various content – partly addresses, partly public lectures on science, articles of a more philosophical character, book-reviews, and so on.

My views have, of course, undergone modifications in the course of time and today I should perhaps no longer express myself in the same way on all matters. Nevertheless I have left everything unchanged, because each piece clearly can and should give nothing but a picture of my views at that time.

The dedication above is not a piece of phrase-making. My thanks for the loftiest spiritual elevation must go to the works of Goethe (his *Faust* is perhaps the greatest of all the products of art and from it I took the mottoes of my first books); likewise to the works of Shakespeare, and so on. But with Schiller the case is otherwise. It is through Schiller that I have come into being. Without him there might have been a man with a beard and nose of the same shape, but I should not have existed.

If anyone else has had on me an influence of the same order of magnitude, it is surely Beethoven. But is it not significant that he, in his greatest work, leaves the last word to Schiller, – and not to the mature Schiller but to Schiller brimming over with the enthusiasm of youth?

Vienna, 8 June 1905   LUDWIG BOLTZMANN

ON THE METHODS OF THEORETICAL PHYSICS*

When the editors of the catalogue of models (published in 1892 for the Association of German Mathematicians) asked me to write on this topic, I soon saw that little remains to be said that is new; so much solid comment on it has appeared just recently. Indeed, our period is marked by an almost excessive critique of the methods of scientific research; one is tempted to call it a critique of pure reason raised to a higher power, if the phrase were not perhaps rather too immodest. Nor can it be my aim further to criticize this critique; I merely wish to make a few clarifying remarks for those who stand remote from these questions but nevertheless are interested in them.

In mathematics and geometry the return from purely analytic to constructive methods and illustration by means of models was at first occasioned by a need for economy of effort. Although this need seems to be purely practical and obvious, it is just here that we are in an area where a whole new kind of methodological speculations has grown up which were given most precise and ingenious expression by Mach, who states straight out that the aim of all science is only economy of effort.

With almost equal justice one might declare that since in business the greatest economy is desirable, the latter is simply the aim of shops and money, which in a sense would be true. However, when the distances, motions, size and physico-chemical properties of the fixed stars are investigated, or when microscopes are invented and with their help the causes of diseases are discovered, we should hardly wish to call this mere economy.

Still, in the end it is a matter of definition what we regard as task and what as means for accomplishing it. Indeed, it depends on the definition of existence whether bodies, or their kinetic energy or even their qualities exist, so that one day we might well simply define away our own existence.

However, enough of this; there is a need for making the utmost use of

* *Populäre Schriften*, Essay 1. First published in *Katalog mathematischer und mathematisch-physikalischer Modelle, etc.*, Munich, 1892.

what powers of perception we possess, and since the eye allows us to take in the greatest store of facts at once (significantly enough we say 'survey'), this gives rise to the need to represent the results of calculations and that not only for the imagination but visibly for the eye and palpably for the hand, with cardboard and plaster.

How little used to be done about this in my student days! Mathematical instruments were almost unknown and physical experiments were often carried out in such a way that no one but the lecturer could see any of it. Since on top of it I was shortsighted and so could not see the writing and diagrams on the blackboard, my powers of imagination were kept constantly on the alert. I was almost going to say fortunately; that, however, would run counter to the purpose of this catalogue review, which must surely be to praise the infinite arsenal of models in mathematics today; besides, the remark would be inaccurate. For although my imaginative gifts did profit, this was only at the expense of the scope of knowledge acquired. At that time the theory of surfaces of the second degree was still the summit of geometrical knowledge and to illustrate it an egg, a napkin-ring and a saddle were enough. What abundance of figures, singularities and shapes evolving from each other the geometer of today must impress on his mind, and what valuable help he receives in this from plaster casts, models with fixed and movable strings, rails and joints of all kinds.

Alongside this, other machines are steadily gaining ground, not for the purpose of representation but in order to take man's place as regards the labour of actually carrying out numerical operations, from the four basic ones to the most complicated integrations.

It goes without saying that both types of apparatus are widely used by physicists who are in any case accustomed to handling instruments. All mechanical models, optical wave surfaces, thermodynamic surfaces in plaster, all kinds of wave machines, devices for representing the refraction of light and other laws of nature are examples of models of the first type. As to constructing apparatus of the second type, we have gone so far as to try to determine the solutions of differential equations that equally hold for an inaccessible phenomenon like friction in gases and an easily measurable one like the distribution of electric current in a conductor of suitably chosen shape, by simply observing the latter and using the values thus read off to calculate the frictional constant for the former. We may likewise recall Lord Kelvin's evaluation by graphical means of the series and integrals that occur in the theory of tides, electrodynamics and so on. Indeed, in his lectures on molecular dynamics he put forward the idea of a mathematical institute for such calculations.

In theoretical physics there are however models that I should like to classify under a third and special head, since they owe their origin to a special method that is being used more and more in precisely that branch of knowledge. I think this is due to practical needs rather than epistemological speculations, but nevertheless the method often carries a highly philosophic imprint so that we must step anew into the field of epistemology.

On the foundation created by Galileo and Newton it was above all the great Parisian mathematicians who, about the time of the French revolution and later, had worked out a sharply defined method of theoretical physics. Certain mechanical assumptions were made from which by means of the principles of mechanics, now arrived at a kind of geometrical self-evidence, a group of natural phenomena were explained. One was indeed aware that the assumptions could not be called correct with apodeictic certainty, but it seemed up to a point likely that they corresponded exactly with reality; they were therefore called hypotheses. Thus matter, the luminous aether required for the explanation of the phenomena of light, and the two electric fluids were all thought of as sums of mathematical points. Between any two such points a force was considered to be acting, whose direction lay in the line joining them while its intensity was an as yet undetermined function of the distance (Boscovich). A spirit who knew all the initial positions and velocities of all these particles as well as all the forces, being able besides to solve all the resulting differential equations, could calculate in advance the whole course of the universe, just as an astronomer can an eclipse (Laplace). One had no qualms in treating these forces, which were taken as originally given and not explicable further, as the cause of phenomena, and their calculation from the differential equations as the explanation of the phenomena.

Later, there was added the hypothesis that, even in bodies at rest, these particles are in motion, which produces thermal phenomena. Particularly for gases, the nature of these motions was very precisely defined (Clausius), and their theory led to surprising mathematical predictions, such as that the frictional constant is independent of pressure, certain relations between friction, diffusion and thermal conduction and so on (Maxwell).

The totality of these methods was so successful that the task of science was taken to be precisely the explanation of natural phenomena. Likewise, the sciences formerly called descriptive began to triumph when Darwin's hypothesis allowed them not only to describe but also to explain biological forms and phenomena. Almost at the same time physics oddly enough took a turn in the opposite direction.

It was above all Kirchhoff who doubted whether the privileged position given to forces viewed as causes of phenomena was justified: whether, with Kepler, we specify the shape of a planetary orbit, indicating the velocity at every point, or, with Newton, we give the force at any point, both are really no more than different ways of describing the facts; Newton's merit is merely the discovery that the description of the motion of celestial bodies becomes especially simple, if we specify the second differential coefficient of their co-ordinates with respect to time (acceleration, force). In half a page, forces had been defined away and banished from nature and physics made into a descriptive science properly speaking. The edifice of mechanics was too solid for this external change to occasion any essential influence on its inner character. Theories of elasticity that did without the idea of molecules were likewise older (Stokes, Lamé, Clebsch). However, in the development of other branches of physics (electrodynamics, theory of pyro- and piezo-electricity and so on), much influence was gained by the view that it could not be the task of theory to see through the mechanism of nature, but only to set up the simplest possible differential equations starting from the simplest possible assumptions (that certain quantities are linear or other simple functions and so on), such that from them we can calculate the phenomena of nature as accurately as possible; as Hertz puts it rather characteristically, the task is merely to represent directly observed phenomena in bare equations, without the colourful wrappings of hypotheses that our imagination lends them. Meanwhile several scientists had already attacked the old system of force-centres and forces at a distance from another even more sensitive direction; one could say from the opposite direction, because they were especially fond of the colourful wrappings of mechanical representation, or from a neighbouring direction because they too renounced recognition of a mechanism at the basis of phenomena, seeing in their own excogitated mechanisms not those of nature but merely pictures or analogies.[1] Several scientists, amongst whom Faraday is foremost, had fashioned for themselves a quite different representation of nature. Whereas the old system regarded force centres as the only reality while treating forces as mathematical concepts, Faraday saw these latter as clearly operative from point to point of the intervening space; the potential function, previously a mere formula facilitating calculation, he regarded as the really existing link in space and cause of the action of forces. Faraday's ideas were much less clear than the earlier hypotheses that had mathematical precision, and many a mathematician of the old school placed little value on Faraday's theories, without however reaching equally great discoveries by means of his own clearer notions.

Soon people tried, especially in England, to attain as illustrative and tangible representations as possible for concepts and notions that till then had played a role only in mathematical analysis. From these endeavours arose the graphic representation of the basic concepts of mechanics in Maxwell's *Matter and Motion*, the geometrical representation of two superimposed sine motions, and all the illustrations occasioned by quaternion theory, such as the geometric interpretation of the operator [2]

$$\Delta = \frac{d^2}{dx^2} + \frac{d^2}{dy^2} + \frac{d^2}{dz^2}.$$

A further circumstance was involved. Most surprising and far-reaching analogies revealed themselves between apparently quite disparate natural processes. It seemed that nature had built the most various things on exactly the same pattern; or, in the dry words of the analyst, the same differential equations hold for the most various phenomena. Thus thermal conduction, diffusion and the distribution of charge in electric conductors follow the same laws. The same equations may be regarded as the solution of a problem in hydrodynamics and in potential theory. The theory of fluid vortices and that of friction in gases show the most surprising analogy with electrodynamics and so on. (See also Maxwell, *Scientific Papers*, Vol. I, p. 156.)

Such influences, too, from the outset pushed Maxwell into a different path when he undertook the mathematical elaboration of Faraday's representations. Already J. J. Thomson (*Mathematical and Physical Papers*, Vol. I) had stressed a series of analogies between problems in the theory of elasticity and in electro-magnetism. In his very first paper on the theory of electricity ('On Faraday's lines of force', see *Scientific Papers*, Vol. I, p.

157), Maxwell declares that he does not intend to propose a theory of electricity; that is, he does not himself believe in the reality of the incompressible fluids and resistances that he is assuming, but merely wishes to give a mechanical example that shows much analogy with electric phenomena, which he wants to present in a form that makes them most readily understandable. In his second paper ('On physical lines of force', *Scientific Papers*, Vol. I, p. 451) he goes much further still, constructing from fluid vortices and friction rollers moving inside cells with elastic walls an admirable mechanism that serves as mechanical model for electromagnetism. Naturally, this mechanism was derided by those who like Zöllner regarded it as a hypothesis in the old sense and thought that Maxwell ascribed reality to it (which in fact he resolutely rejects, and merely expresses the modest hope that by means of such mechanical fictions further research in the theory of electricity will be more helped than hindered). And they were indeed helped, for through his model Maxwell reached those equations whose peculiar and almost inconceivably fantastic powers were described so vividly by the man best qualified to do so, namely Heinrich Hertz (in his lecture on the relation between light and electricity, published in Bonn, 1890). To this I wish to add only that Maxwell's formulae were merely consequences of his mechanical models, so that Hertz's enthusiastic praise is due in the first place not to Maxwell's analysis but to his ingenuity in discovering mechanical analogies.

Only in Maxwell's third important paper ('A dynamical theory of the electromagnetic field', *Scientific Papers*, Vol. I, p. 526) and in his *Treatise on Electricity and Magnetism* (Oxford, 1881) do the formulae become more detached from the models, a process that was subsequently completed by Heaviside, Poynting, Rowland, Hertz and Cohn. Maxwell still uses mechanical analogies, or, to use his term, dynamic illustrations. However, he no longer specifies them in detail but rather looks for the most general mechanical assumptions that are apt to lead to phenomena analogous to those of electromagnetism. Thus Thomson, by extending his earlier mentioned ideas, was led to the quasi-elastic and quasi-unstable aether and its illustration by means of a gyrostatic-dynamic model.

Naturally Maxwell transferred the same mode of treatment to other branches of theoretical physics as well. For example, his gas molecules repelling each other with forces proportional to the inverse fifth power are to be taken as a mechanical analogy, and there has been no lack of recent critics who misunderstood him and declared his hypothesis to be as unlikely as absurd.

Gradually, however, the new ideas gained admission in all fields. From thermodynamics I here mention only Helmholtz's famous dissertations on the mechanical analogies of the second law. Indeed, it turned out that they were more in tune with the spirit of science than the old hypotheses, besides being more convenient for the scientist himself. For the old hypotheses could be upheld only so long as everything went well; but now the occasional lack of agreement was no longer harmful, for one cannot reproach a mere analogy for being lame in some respects. Hence the old theories, such as the elastic theory of light, gas theory, the chemist's benzene rings and so on, were soon taken as no more than mechanical analogies.

In the end, philosophy generalised Maxwell's ideas to the point of maintaining that knowledge itself is nothing else than the finding of analogies. This once again meant that the old scientific method had been defined away and science spoke merely in similes.

At first, of course, all these mechanical models existed only in thought, they were dynamic illustrations in imagination, nor could they be carried out in practice at this general level. Yet their great importance stimulated people to construct at least their basic types.

In Part 2 of this catalogue there is a report on such an attempt undertaken by Maxwell himself and another by the present writer. Fitzgerald's model, too, is currently at the Nurnberg exhibition as well as Bjerknes's model, which we owe to similar tendencies. Further models that belong here are those constructed by Oliver Lodge, Lord Rayleigh and others.

They all show how the new approach compensates the abandonment of complete congruence with nature by the correspondingly more striking appearance of the points of similarity. No doubt the future belongs to this new method; but just as it was wrong earlier to regard the old method alone as correct, so it would be one-sided now to take it as quite worn out in spite of all it has done, and not to cultivate it alongside the new one.

NOTES

[1] Compare the almost ethereally structured and crystal clear though colourless theory of elasticity in Kirchhoff's lectures with the crudely realistic account in Vol. 3 of Thomson's *Mathematical and Physical Papers* which concerns not ideally elastic bodies

but steel, rubber, glue; or with the often childlike naiveté of Maxwell's language, who in the midst of formulae mentions a really effective method for removing fat stains.
[2] Maxwell, *Treatise on Electricity and Magnetism* 1873, Vol. 1, Sec. 29: Nature of the operator ∇ and ∇^2. This was later noticed by others too: Mach, *Wien. Sitzungsber.* **86** (1882) 8. Cf also *Wied. Beibl.* **7**, 10; *Comptes Rendus Ac. de Paris* **95**, 479. [*Editor's note*: Consistency with modern practice (and with Boltzmann's own in his *Vorlesungen über die Principe der Mechanik*) would demand '∂' for 'd' throughout. For 'Δ' it is now more common to use Maxwell's won '∇^2.]

THE SECOND LAW OF THERMODYNAMICS*

When it came to be my turn to speak on a solemn occasion to this gathering, attended by so many to whom I owe my scientific education, I was well aware how difficult was the honourable duty I had undertaken and only reluctantly began to shoulder it. Forgive me, therefore, if I feel I must devote some words of apology to the very choice of my subject. This choice is no doubt easier for the philosopher and historian who remain in constant touch with the general public. In natural science it used often to be the custom to discuss more general topics of so-called philosophic or metaphysical interest. If today I depart from this custom, I certainly do not wish to provoke the suspicion that these more general questions seem to me insignificant or unimportant as against the countless special problems raised by contemporary science. It is only the manner in which they have been treated to date, in some cases one almost feels tempted to say the fact that they are treated at all at this time, that seems to me mistaken. Hence the peculiar phenomenon that, while in special fields effort is often amply repaid, as regards general questions the most strenuous attempts are often unattended by any success: in the former case, for all the controversy as to detail there is in the main agreement, while in the latter the most contradictory views find their supporters and those who worked together in harmony on special questions no longer understand each other.

Nowhere less than in natural science does the proposition that the straight path is the shortest turn out to be true. If a general intends to conquer a hostile city, he will not consult his map for the shortest road leading there; rather he will be forced to make the most various detours, every hamlet, even if quite off the path, will become a valuable point of leverage for him, if only he can take it; impregnable places he will isolate. Likewise, the scientist asks not what are the currently most important questions, but "which are at present solvable?" or sometimes merely "in

* *Populäre Schriften*, Essay 3. Address to a formal meeting of the Imperial Academy of Science, 29 May 1886.

which can we make some small but genuine advance?". As long as the alchemists merely sought the philosopher's stone and aimed at finding the art of making gold, all their endeavours were fruitless; it was only when people restricted themselves to seemingly less valuable questions that they created chemistry. Thus natural science appears completely to lose from sight the large and general questions; but all the more splendid is the success when, groping in the thicket of special questions, we suddenly find a small opening that allows a hitherto undreamt of outlook on the whole.

Galileo's inclined plane, Stevin's chain have become mighty points of support from which mechanics penetrates not only into the external relations between bodies but also into the nature of matter and energy. The remarkable facts that chemists daily discover are as many new proofs of atomism. Joule's experiments have definitively decided the old controversies about the nature of work, impulse and *vis viva*. The great query: whence do we come, whither will we go, has been discussed for thousands of years by the greatest of men of genius and turned this way and that with immense ingenuity; whether with any measure of success I do not know, but certainly without any essential and undeniable progress: that was not attained until the present century, thanks to most careful studies and comparative experiments on the breeding of pigeons and other domestic animals, on the colouring of flying and swimming animals, by means of researches into the striking similarity of harmless to poisonous animals, through arduous comparisons of the shape of flowers with that of the insects that fertilize them. All these are indeed fields of apparently subordinate importance, but in them genuine success could be attained which precisely became the solid operational basis for a campaign into the realm of metaphysics attended by success that is unique in the history of science.

Schiller remarked on the inquirers of his time: "to capture truth they sally forth with nets and staves, but with the steps of a master-spirit she sweeps through their midst". How much more, seeing today's weapons of physics and chemistry, would he have doubted whether with such a chaos of apparatus truth could be captured, and the same picture meets us in the present day work-places of mineralogists, botanists, zoologists, physiologists and so on. In such pieces of apparatus not only do I see devices for harnessing the forces of nature in new ways, but indeed I view them with much greater reverence, and venture to say that I regard them as the true means for unveiling the nature of things. In this, of course, many problems are like the question once put to a painter, what picture he was hiding behind the curtain, to which he replied "the curtain *is* the picture". For when requested to deceive experts by his art, he had painted a picture representing a curtain. Is not perhaps the veil that conceals the nature of things from us just like that painted curtain?

If we regard the apparatus of experimental natural science as tools for obtaining practical gain, we can certainly not deny it success. Unimagined results have been achieved, things that the fancy of our forebears dreamt in their fairy tales, outdone by the marvels that science in concert with technology has realised before our astonished eyes. By facilitating the traffic of men, things and ideas, it helped to raise and spread civilisation in a way that in earlier centuries is parallelled most nearly by the invention of the art of printing. And who is to set a term to the forward stride of the human spirit! The invention of a dirigible airship is hardly more than a question of time. Nevertheless I think that it is not these achievements that will put their stamp on our century: if you ask me for my innermost conviction whether it will one day be called the century of iron, or steam, or electricity, I answer without qualms that it will be named the century of the mechanical view of nature, of Darwin.

After this confession you will take it with more tolerance if I am so bold as to claim your attention for a quite trifling and narrowly circumscribed question, nor will you accuse me of disregarding large and general questions if I turn to things that are as yet unrelated to them. In any case, treating a narrowly circumscribed specialist topic before a wider public should not be entirely without interest. Indeed, the time is long past when a mortal could encompass all or even a fair number of branches of science; today we must limit ourselves not just to one definite branch but to a smallish part of one. At the same time, however, the various branches interpenetrate more and more, so that in spite of extreme division of labour no individual must ever lost sight of other fields which, unfortunately, is not possible without occasional at least hurried glances at their detail.

One used to divide the totality of natural sciences into two main groups: one was called that of descriptive sciences; the other, which includes physics, chemistry, astronomy, physiology and, insofar as they were counted as sciences, mathematics, geometry and mechanics, would then have to be called the group of explanatory sciences. We must not be surprised that the disciplines of natural history have long since protested against the title 'descriptive' that so greatly limits their task. Since the mighty upswing of geology, physiology and so on, but above all since the general acceptance of the ideas of Darwin, these sciences boldly undertake to explain the forms of minerals and of organic life. However, it is strange to see that on the other side the opposite turn is being taken almost at the same time. In this comprehensive work on mechanics, Kirchhoff very clearly sets himself as a task merely to describe natural phenomena as simply and perspicuously as possible, renouncing all explanation, and since then what in physics used to be called explanation has repeatedly been called a mere description of the facts; this because one wishes to avoid a vagueness in the concept of explanation. If one seeks to explain motions from forces and these from the essence of things, that is phenomena from things in themselves, one always seems to start from the view that explanation requires reducing the explicandum to some quite new principle external to it. This view is alien to natural science, which merely resolves complex things into components that are simpler but the same in kind, or reduces complicated laws to more fundamental ones. If now this process is often successful it becomes so much a habit that we have no wish to stop even at its natural end. Usually one even regards it as a limitation of our intellect that, assuming we had succeeded in finding the simplest basic laws, we could then not explain or ground them further, that is resolve them into simpler ones; that as regards the existence of the most elementary constituents we are in any case unable to comprehend them, that is reduce them to simpler ones still. Are we not here once more placed in front of that painted curtain mentioned earlier? Are we going to regard it as a limitation of our sense of sight that nobody can tell what picture is concealed behind the curtain? We shall be able to retain the word 'explain' if from the outset all such reservations are kept at a distance.

We infer the existence of things only from the impressions they make on our senses. It is thus one of the fairest triumphs of science if we succeed in inferring the existence of a large group of things that mostly escape our sense perception; thus astronomers, from often slight traces of light, were able with near certainty to infer the existence of many celestial bodies many of them a thousand and even a million times bigger than the earth and at distances that make the mind reel merely to think of them. If there-

fore amongst the intruments to which metaphysics owes gratitude I failed
to mention those of astronomical observations, from the simplest dioptric
devices of the ancient Egyptians to the telescopes of Galileo and Kepler
and the giant instruments of Clark, this merely proves how incomplete
was my list. What astronomy has succeeded in doing for phenomena on
the largest scale has similarly been achieved for the smallest dimensions.
All observation points to things so small that only millions together can
excite our senses. We call these things atoms and molecules. Conditions
in the investigation of atoms are in many ways much less favourable still
than in astronomy. We can always think of celestial bodies as being similar to our earth; even if as regards size, state of aggregation, temperature,
and so on, there are bound to be the most varied differences, we can still
think of a mass of molten metal or large glowing spheres of gas, for which
spectral analysis offers further clues. About the constitution of atoms,
however, we know as yet nothing and will continue to do so until we
succeed in formulating a hypothesis from the facts observable by the
senses. Strangely enough the first success is here again to be expected from
the art that had proved so powerful in the investigation of celestial bodies
too, namely spectral analysis. That there are such tiny individual things
whose joint effect first forms bodies perceivable by the senses is of course
only a hypothesis; just as it is only a hypothesis that what we see in the sky
is caused by celestial bodies of such size and distance, as indeed it is basically only a hypothesis that besides myself there are other human beings
that feel pleasure and pain, that there further exist animals, plants and
mineral bodies. Perhaps one day a hypothesis that the stars are mere
sparks of light will explain celestial phenomena better than our current
astronomy, but it is unlikely. Perhaps the atomistic hypothesis will one
day be displaced by some other but it is unlikely.

This is not the place for naming all the reasons that might be advanced
for this. There will be no need to recall the ingenious inferences of Thomson who used the most varied methods to work out with quite satisfactory agreement how many of these individual things make up a cubic
millimetre of water. I need not mention that, besides many facts of chemistry, it was by means of the atomistic hypothesis that science succeeded
in calculating in advance the temperature dependence of the frictional
constant for gases, the absolute and relative constants of diffusion and
thermal conduction, which can surely be put alongside Leverrier's calcula-

18   FROM 'POPULÄRE SCHRIFTEN'

tion establishing the existence of the planet Neptune or Hamilton's prediction of conical refraction. A detailed discussion of these two problems will here be the less required as each is for ever linked with the name of a member of this academy. Let me merely make a brief comment on the first calculation of the frictional constant by Maxwell.

From his theory he derives the result that for a whole class of phenomena the resistance to a body moving in a gas does not depend on the density of the gas. These phenomena are marked by the fact that for them the mass of gas plays no part in comparison with that of the moving bodies. All previous observations pointed against it, resistance had always been found to be much greater in dense air than in thin. Besides the result seemed to be a priori unlikely, for if resistance were independent of density, it would have to remain the same at zero density, when there would be no gas at all. All this could not have escaped Maxwell and when he first published his result he confessed that he almost preferred to believe that his calculations were faulty than that such absurd results were correct. Since then many cases belonging to this class of phenomena were carefully examined and the only thing exposed as false was Maxwell's lack of trust in the power of his own weapons. There remains no doubt that in these cases resistance is really independent of density within a wide range. If density becomes too small, resistance finally drops and vanishes where no gas remains, but here too theory was able to predict, with numerical precision, within what limits Maxwell's law was valid.

Closely connected with atomism is the hypothesis that those elements of bodies are not at rest, forming matter by lying rigidly alongside each other, like bricks in a wall, but that they are in vigorous motion. This hypothesis, called the mechanical theory of heat, is likewise a view solidly based on facts. Its numerical formulation derives from the principle of energy first clearly enunciated by Robert Mayer. Energy may take three forms, the visible motion of bodies, thermal motion, that is the motion of the smallest particles, and finally work, that is the separation of mutually attracting bodies or the approach of repelling ones. This last form seems less comprehensible; a hint is given by the circumstances of work as regards magnets and electric currents: these depend so much on configuration that we spontaneously form the notion of other motions intervening that are not only invisible like thermal vibrations of molecules but even so far undefined by any hypothesis, for example motions of an as yet unknown medium of the luminiferous aether. When repelling bodies approach or attracting ones separate, the motions in this medium ought to increase; small wonder therefore that in return the sum of visible and thermal motions becomes smaller, since part of it goes over into the hypothetical medium. The reverse would hold for the opposite case. Thus we could easily deduce all phenomena from a general principle. Heat, visible kinetic energy and work could be produced from or transformed into each other while their quantity remains always the same.

However, alongside this general principle the mechanical theory of heat has placed a second one that limits the first in a rather unsatisfying way, the so-called second law of thermodynamics, expressed roughly as follows: work and visible kinetic energy can be transformed into each other and into heat unconditionally, but conversely the reversion of heat into work or visible kinetic energy can occur either not at all or at best only in part. If in this form the principle looks already like an uncomfortable appendix to the first one, it becomes much more inconvenient still by its consequences. The energy form we need is always that of work or visible motion. Purely thermal vibrations slip through our hands and escape our senses and are for us synonymous with rest; hence the thermal form of energy used often to be called dissipated or degraded energy, so that the second law proclaims a steady degradation of energy until all tensions that might still perform work and all visible motions in the universe would have to cease.

All attempts at saving the universe from this thermal death have been unsuccessful, and to avoid raising hopes I cannot fulfil, let me say at once that I too shall here refrain from making such attempts.

Rather my intention is merely to examine the second law a little more closely from another angle. Molecular thermal motions are most probably such that a given state of motion is not shared by a large group of neighbouring molecules, but that in spite of constant mutual influence each molecule pursues its own independent path, appearing as it were as an autonomously acting individual. One might therefore think that this autonomy of the parts would at once have to show itself in the external properties of bodies; for example, that in a horizontal metal bar now the right and now the left end must become spontaneously hotter according as the molecules happen to vibrate more intensely at one or the other place, or that if in a gas a large number of molecules happen to be moving towards the same point at the same time, a sudden increase in density must occur there. However, we observe none of this, and the reason why this is so is nothing other than the so-called law of large numbers.

As is well known, Buckle has shown by statistics that if only we take a large enough number of people, then so long as external circumstances do not change significantly, there is complete constancy not only in the processes determined by nature, such as number of deaths, diseases and so on, but also of the relative number of so-called voluntary actions such as marriage at a certain age, crime, suicide and the like. Likewise with molecules: the pressure on a piston arises from various molecules impinging, now more now less strongly, now head on now at an angle, but because of the large number of colliding molecules not only does the total pressure remain constant but also the same average intensity of collisions falls on any visible or observable part however small. If we notice that the pressure is bigger at any point, we shall at once look for an external cause that moves the molecules to flow preferentially to that point. If now in a given system of bodies there is a given amount of energy, the latter will not arbitrarily transform itself now in one now in another manner, but it will always go from a less to a more probable form; if the initial distribution amongst the bodies did not correspond to the laws of probability, it will tend increasingly to become so. Precisely those forms of energy that we wish to realise in practice are however always improbable. For example, we desire that the body move as a whole; this requires all its molecules to have the same speed in the same direction. If we view molecules as independent individuals, this is however the most improbable case conceivable. It is well known how difficult it is to bring even a moderately large number of independent individuals to do exactly the same in exactly the same manner. Yet only if all motions thus agree can we attain the highest aim of unconditional transformation. Every deviation from agreement amounts to degradation of energy. Equally improbable is the energy form of pure mechanical work, whereas in chemical work a mixture of atoms can occur that corresponds at least in part to the laws of probability.

Therefore, what previously we called degraded energy forms will be none but the most probable forms; or, better, it will be energy that is distributed amongst the molecules in the most probable way. Imagine a number of white balls into which a different number of black but otherwise identical balls is introduced. At the start let one place be occupied only by white, another only by black balls. If however we mix them by hand or expose them to some other influence that constantly alters their relative position, then after some time we will find them variously mixed together. Just so it is with a hot body that is hotter than its surroundings: indeed, there we have a fairly large group of rapidly moving molecules amongst groups that move more slowly. If we bring the hot body into direct contact with its colder surroundings, there develops a velocity distribution that corresponds to the laws of probability. The temperature becomes equalised. If however we adopt detours, we can use the existing improbabilities of energy distribution to produce other improbable energy forms that would not develop spontaneously. When heat goes from a hotter to a colder body we can transform part of the transmitted heat into visible motion or work, as happens for instance in steam engines and any other heat engines.

The same thing will be possible whenever the initial energy distribution does not correspond to the laws of probability, for example when a body is colder than its surroundings, or when at one point of a gas the molecules are more crowded and at another less, and so on. Suppose in the lower half of a container we had pure nitrogen and in the upper pure hydrogen, both at the same pressure and temperature, this distribution would not correspond to the law of probability, which requires that all molecules be uniformly mixed, like the white and black balls above. If the gases mix directly, this is analogous to the case that between two unequally hot bodies the temperature adjusts itself directly, which likewise occasions no transformation of heat into work. It is however conceivable that the mixing occurs via detours and that part of the heat contained in the two gases is changed into visible motion or work. And indeed, as Lord Rayleigh was the first to show, this can in fact be realized.

In a single gas not all molecules will have the same speed, but some rather greater and others less than average, and it was Maxwell who first proved that the various speeds have the same distribution as the errors of observation that always creep in when the same quantity is repeatedly determined by measurement under the same conditions. That these two laws agree cannot, of course, be taken as accidental, since both are determined by the same laws of probability. If one could produce a gas in which all molecules had exactly the same speed, this too would be an energy distribution greatly deviant from the most probable one. If therefore this form of energy could so far never be produced in practice, we could in any case maintain a priori that its transition into ordinary heat would likewise occasion the production of improbable energy forms, no differently from the transition of heat from a hotter to a colder body.

Now we are able not only to declare qualitatively that one energy distribution is quite improbable and another probable, but the calculus of probability as in all other cases that fall under it enables one to set up the precise measure of probability of any energy distribution, provided of course that the mechanical conditions of the system are known. (For the logical foundations of the calculus, see the treatise on the calculus of probability by Kries)[1]. To every energy distribution therefore corresponds a quantitatively determinable probability. Since in the most important cases for practical purposes this is identical with the quantity that Clausius calls entropy, we too shall use this name. All changes in which entropy increases will occur spontaneously, as Clausius puts it. On the other hand, entropy can diminish only if in return some other system gains the same or a greater amount of it. If initially we have two bodies of different temperatures, which then equalise, the probability of the initial state with different temperatures and of the final one which is the more probable can be calculated exactly so that we can determine how much of the heat transferred can be turned into work; only if the initial temperatures were markedly different, as that of burning coal or oxy-hydrogen gas relatively to our ordinary surroundings, can the transferred heat be almost completely transformed into work. In mathematics one usually puts it thus: if the temperature drops from infinite to finite, almost all heat transferred can be transformed into work; the infinitely higher temperature is, as it were, infinitely improbable. Similarly, the case of all atoms moving at the same speed in the same direction, that is a body undergoing visible loco-motion, corresponds to an infinitely improbable configuration of energy: visible motion behaves like heat of infinitely high temperature, it can be completely transformed into work.

A machine is a device using available power to overcome a load. One always works out the case where the applied effort just balances the load, though this is as yet of no practical use; as long as equilibrium prevails, the load cannot be moved a hair's breadth, that would require a small increment to the effort. Just so we proceed in thermodynamics: one always considers such energy transformations as leave the probability of energy distribution always constant; these we call reversible changes of state, for since the probability remains always the same, the change might just as well run in the opposite sense. Strictly speaking, of course, they can run in neither, just as little as the effort at equilibrium can move the load, since energy transformation can actually occur only if this makes the state of the system more probable. However, if the difference in probability is taken as very small, one can come arbitrarily close to reversible changes. In this sense the thermodynamicist thinks of heat as travelling from one body to another at exactly the same temperature, or of a piston as moving when pressure and counter pressure are exactly equal; in practice, the second body will always have to be a little colder and the counter pressure a little smaller. Reversible changes have been imagined with the most various bodies in the most various ways. They always led to remarkable relations between properties, whose connections would not otherwise have been suspected. Insofar as these relations have been tested by experiment, they have regularly proved correct. Thus were discovered the relation between the specific heats, the moduli of compression and thermal expansion, the change of specific volume at solidification and the change with pressure of the freezing point, the supersaturation of vapours by expansion and their other properties, the solubility of salts, their specific weights and the vapour pressures of their solutions, between magnetic and thermal properties of bodies, between heats of formation, electromotive force and its dependence on temperature.

The sun has been extolled as the energy source not only of animal and plant life and meteorological processes, but of all terrestrial work processes, except the sea mills of Agrostoli.

Helmholtz has shown that the heat originating from anthracite is only stored solar heat, but I do not know whether it has been pointed out clearly enough why just this source of energy is of such great use; in the bodies on the earth's surface that are immediately accessible to us, an amount of energy is stored of whose size we have not the least conception. If the heat produced by the Niagara falls is already enough to drive a considerable proportion of all our machines, what inexhaustible supplies of energy would be at our disposal if we could transform all the heat contained in the bodies of our surroundings into work. Yet this cannot be done, because the energy they contain, except insofar as the sun might produce differences in temperature, is already almost in the most probable distribution, so that any attempt at distributing it in ways more suitable to our ends will fail. Between sun and earth, however, there is a colossal temperature difference; between these two bodies energy is thus not at all distributed according to the laws of probability. The equalisation of temperature, based on the tendency towards greater probability, takes millions of years, because the two bodies are so large and far apart. The intermediate forms assumed by solar energy, until it falls to terrestrial temperatures, can be fairly improbable, so that we can easily use the transition of heat from sun to earth for the performance of work, like the transition of water from the boiler to the cooling installation. The general struggle for existence of animate beings is therefore not a struggle for raw materials – these, for organisms, are air, water and soil, all abundantly available – nor for energy which exists in plenty in any body in the form of heat (albeit unfortunately not transformable), but a struggle for entropy, which becomes available through the transition of energy from the hot sun to the cold earth. In order to exploit this transition as much as possible, plants spread their immense surface of leaves and force the sun's energy, before it falls to the earth's temperature, to perform in ways as yet unexplored certain chemical syntheses of which no one in our laboratories has so far the least idea. The products of this chemical kitchen constitute the object of struggle of the animal world.

Lest I become involved too much in details that can interest only the expert, I must here refrain from undertaking the task, however attractive it might be, of citing further special cases in order to clarify in what way the energy distribution in a system of bodies assumes ever more probable forms, or of giving examples in order to illustrate what kind of detours enable us to produce fairly improbable distributions by starting from ones that are even less so but still given in nature and artificially guiding them into the desired paths. Let me however touch on just one area of somewhat general importance; you may have noticed that on various occasions I spoke not of bodies in general but only of gases. The reason for this lies in the fact that the molecules of gases are so far apart that they no longer exert significant forces on one another; since the external forces acting on gases can usually be neglected, their molecules are indeed precisely in the state of the black and white balls mentioned earlier. Their mixture according to the laws of probability is not disturbed by alien influences. Every point within the vessel, every direction of motion is for them equally probable. Not so for the various values of the magnitude of velocity. Let the total energy of the gas be given. The greater the speed of a molecule, the more restricted therefore the choice of speed of the remaining ones: hence large speeds for a single molecule are always improbable up to the most extreme and improbable case of a single molecule possessing the whole kinetic energy contained in the gas, all others having none. Every gas molecule is flying about with the speed of a shell from a gun and within one second collides many million times with other molecules. Who then could conceive even approximately what confused turmoil agitates the elements of these bodies! However, the average results can be found by combinatorial analysis as simply as in the case of a game of lotto.

In liquid fluids and solid bodies there is in addition the effect of molecular forces. Indeed, a considerable amount of energy is required to separate a given mass of fluid water into molecules of vapour. One imagines that between water molecules there are attractive forces which of course increase the probability of coalescence.

As indicated above, one might alternatively ascribe these forces to a medium. Separation of two water molecules would have to increase the medium's energy. The relevant mechanism is of course quite unknown to us; however the energy of an ordinary liquid is altered by relative change of position of eddies or solid rings within them. The energy arising in the medium would then be lost for thermal motion. The separation of two water molecules would then be more improbable not because of an attractive force but for the same reason as greater speeds of molecules above. For this separation would lower the thermal energy of the mass of water, thus diminishing the number of possible energy distributions amongst the remainder of molecules.

I can here sketch the final result only in a few strokes. Consider a liquid in a large sealed vessel which it does not fill completely. If the liquid contains very little energy, it can happen that there is not enough to separate even a single molecule: all of them would have to cleave together; even if this situation is never actually realized, is is enough if the whole energy is consumed in separating relatively few molecules for only vanishingly small amounts of vapour to form over the liquid. As the temperature rises this vapour will become progressively denser and the liquid less cohesive. Take now the other extreme case: let the total energy be very large, so that in comparison the small amounts absorbed from or transferred to the medium when two molecules are merged or separated are negligibly small (the work done by the molecular forces will vanish), then the whole mass must behave like a gas both at arbitrarily low or high density. The boundary of the two states is what we call the 'critical temperature': just below it there remain both liquid and vapour, but they differ little, the work of molecular forces no longer counts for much; while above it everything is uniform, one cannot say whether liquid or gaseous, since the two branches run into each other.

If two different fluids are mixed, heat is produced if mutual attraction predominates, but absorbed in the opposite case. It would be incorrect to think that in the former case the fluids mix spontaneously and in the latter not, for a uniform mixture is much more probable than complete separation, as with the often mentioned white and black balls. That is why in gases too a mixture will always occur, although unaccompanied by any noticeable production of heat. If the mixture of liquids produces heat, they will mix the more readily of themselves, but if cooling arises, the greater probability of the mixed state may still be decisive. Only when cohesive forces are significantly preponderant will the tendency towards mixture be overcome.

Much the same occurs in the operation of so-called chemical affinities. It is a fact that when two like or unlike atoms combine in certain ways, very considerable amounts of energy are set free. The so-called valency of atoms alone shows that here it is not only mutual approach to a certain small central distance that is decisive, as we previously imagined it for water molecules, but that this release of energy occurs only when the constituents assume a certain relative position. Take first monovalent atoms, where the effect occurs only when two atoms join, while a third one supervening releases hardly any further energy, e.g. as in gaseous chlorine. At low temperatures the separation of each pair of atoms is highly unlikely, as in the earlier case of the vapour molecule separating from the liquid mass. As the temperature increases more and more pairs of atoms will separate (dissociate), until they are all dissociated. Qualitatively, this hypothesis has long been established amongst chemists, but the calculus of probabilities allows a quantitative determination. The total energy contained in the gas can be determined, so that for the ratio of undissociated to dissociated molecules we obtain a formula with only two unknown constants to be determined from observation: one is the energy released when two chlorine atoms combine into a molecule, the other determines the space within which one atom must be relatively to a second in order that they should appear as chemically joined, what I want to call the binding range. Insofar as we have observations, the formula mentioned agrees with experience.

Accordingly, we must view the dependence of degree of dissociation on pressure as follows: given the number N_1 of dissociated atoms, then whenever a new atom lies within the binding range of one of them a molecule is formed, whenever it lies in the remaining space it is free. If therefore that space is doubled without changing N_1, which in view of the vanishing smallness of the binding ranges amounts to doubling the total volume of the gas, the probability that two atoms are combined drops by half; the bigger we make the volume containing the mass of the gas, the higher becomes the degree of dissociation at constant temperature, given the number N_1 of dissociated atoms, the number N_2 of molecules will be inversely proportional to the volume. Similarly we determine the temperature dependence of the degree of dissociation. I have tried to work out from Victor Meyer's observations on iodine vapour the energy released on the combination of two iodine atoms into a molecule and the binding range. The former comes out at $\frac{3}{4}$ of the heat of combustion of hydrogen, the latter is as yet affected by considerable uncertainty, although it is certainly very small compared with a sphere whose diameter equals the mean distance of two solid iodine atoms. I will not call this the size of an iodine atom to avoid the suspicion that I want to ascribe to atoms any similarity with solid spheres or other tiny solid bodies. For multivalent atoms the task of combinatorial analysis is of course much harder but not insoluble.

To the question when liquids mix of their own accord corresponds in chemistry the principle of Berthelot. That chemical compound which produces most heat is always attended by the greatest probability and will always occur by preference; if the excess of heat is large, it will occur on its own: this last is Berthelot's principle. If however the excess is small, other compounds may always occur as well in small amounts; these are the exceptions to the principle. Consider two kinds of monovalent atoms A and B present in equal numbers, and assume that the heat of combination

(A_2) of two atoms A into a molecule A_2 is exactly equal to the heats of combination (B_2) and (AB), the last one for the formation of a molecule AB from one atom each of A and B. Let the three heats of combination be so large that rather few molecules are dissociated, and that everything is in the gaseous state; then according to the calculus of probability half the molecules will be AB and a quarter each A_2 and B_2, that is the partial pressure of the gas AB is twice that of the gases A_2 and B_2. Just as the probability of drawing two white or two black balls is $\frac{1}{4}$, and one white and one black $\frac{1}{2}$ if two are drawn from a bag containing half of each. If the heat of formation (AB) is less than $\frac{1}{2}((A_2)+(B_2))$, that is the transformation of A_2 and B_2 into two AB has a negative heat of formation, then less of AB will be formed if the two bodies A_2 and B_2 are mixed; nevertheless, in spite of the negative heat of formation a measurable amount of the compound AB may well form itself. Not until the heat of formation becomes sufficiently negative will the quantity of AB formed become undetectable, but so long as the heat of formation is negative the total mass can never completely turn into AB. I must here omit any further discussion of combinations of more than two atoms or applications to real cases, as for example a comparison with the investigations of Rathke (*Nat. Ges. Halle*, Vol. 15, 1881).

How, one might ask, is it that not all the compounds corresponding to the possible combinations of atoms are always formed, since each has a greater or smaller probability in favour of it? About this, too, the calculation informs us, for the quantities formed are given by exponential magnitudes whose gigantic growth or fall has often been illustrated, for example by the vast present value of one penny invested at compound interest at the birth of Christ, or by the tale of the man who invented chess. From the above-mentioned formula one finds that iodine vapour at 30 °C must indeed contain some dissociated atoms, but in 1 000 kg of it their weight would be only one part in a hundred million of a milligram $(10^{-11}$ g). (The fact that iodine is not very volatile at this temperature is here unimportant. No doubt we should find an even more favourable result for chlorine or bromine.) For the same reason, oxy-hydrogen gas at ordinary temperature does not give rise to much water however long the period, even though water is by far the most probable compound, since for one water molecule to form at least one oxygen or hydrogen atom must be dissociated, which certainly occurs incredibly less often than with iodine, by what factor we do not know, since we lack all data concerning oxygen and hydrogen.

Radiant heat inside a perfectly black container at constant temperature corresponds to the laws of probability, but the vibrations of light at the conditions of temperature of the earth's surface are motions of greater regularity and thus constitute a rather improbable energy form, a transitional stage when energy goes from a very hot to a cold body; hence its considerable dissociating effect without much heating.

First I considered a clear case of physical mixture and later a clear case of a chemical compound; however different the characteristic marks of these two extremes, we can imagine a continuous path between them by means of various intermediate steps. By quite different methods, Helmholtz found that water can never be freed from the last trace of dissociated hydrogen and oxygen; only if at the start all atoms were combined into H_2O, would there be just one atom of O for every two of H. Since the former was never so, neither is the latter and the numerical proportions between the dissociated atoms may be quite different; hence no chemical compound will contain with absolute precision the proportion of atoms corresponding to its formula, even if the deviations are millions of times smaller than the quantity of dissociated iodine atoms calculated above. However, we may conceive of transitional bodies where one or the other ingredient might be in considerable excess. Such compounds would not show properties totally different from those of its components; special properties arising from the mixture will merely reach a maximum at some definite proportion of the components and certain groups of atoms will form preferentially though not exclusively, as indeed is strictly speaking not the case in any chemical compound. The more marked this maximum, the more evident the character of the chemical compound; the flatter the maximum, the more evident the character of the physical mixture. Examples are offered by the hydrates of many acids, for instance sulphuric acid, by salts that contain crystallised water, by many metals whose alloys resemble chemical compounds. If such bodies are distilled, or allowed to crystallize, at a certain pressure, one often obtains products of fixed composition which however varies if pressure or other external circumstances vary. Here too nature knows no jumps. How striking is the difference between animal and plant, yet the simplest forms continuously merge into each other, so that some are just on the border and represent animals as readily as plants. In natural history the various species are mostly sharply separate, but here and there continuous transitions occur. No doubt no one would want to abolish the concepts of animal, plant and species on that account, but the question whether a certain form is or is not a new species will occasionally admit of no answer, because of the impossibility of formulating the concept of species with absolute precision. Just as little will anybody ever wish to abolish the concept of chemical compound or even need essentially to alter its use, yet in individual cases the question whether we are dealing with a chemical compound or a mixture will be void because a sharp definition of the concept is impossible. Indeed, if these principles proved themselves as correct, they would exclude from the outset many assumptions made by chemists, for example that of Rüdorff (see Lothar Meyer, *Moderne Theorien der Chemie,* p. 236) declaring that sodium chloride is free of water above −9 °C but contains two chemically bound water molecules below that temperature. Such a sudden change in composition at a given temperature would be impossible in a simple separation of a molecule into two smaller ones. At least there would have to be a finite temperature interval of gradual dissociation in between. Nor, as Bineau thought, could sulphur molecules suddenly change from six to two atoms.

Since a given system can never of its own accord go over into another equally probable state but only into a more probable one, it is likewise impossible to construct a system of bodies that after traversing various states returns periodically to its original state, that is a perpetual motion machine. And so we have arrived where one usually begins when considering the second law. One sets up the axiom that it is impossible to construct a perpetuum mobile from a finite number of bodies. The axiom is formulated in equations that are called the fundamental equations of the second law, and then one is amazed that, on the assumption that the world is a large system with a finite number of bodies, it follows from these equations that not even the world as a whole can be a perpetuum mobile, which was really contained already in the assumption. However enticing such vistas of the universe and however suggestive they doubtless often prove, I still think that here we just extend empirical propositions far beyond their natural boundaries.

Since atoms have given such faithful service in all branches of physics and chemistry, the question arises whether it is likely or even possible that we can by their means explain the phenomena of animal life as well, that is thought and sensation. I do not know whom immediate consciousness tells beyond doubt today as once it did Herbart that the 'ego' is a simple entity; but sensations, the elements of our whole thinking, surely those are something simple? I think about this too our consciousness cannot tell us anything, for it leaves sensation completely undefined, telling us only that one of red is different from one of blue but not whether both are simple elements or complicated dislocations of countless atoms perhaps comparable to a wave motion. We can sense red, but what a sensation might be we cannot sense.

Perhaps it goes against the grain to regard what was presented to us even before we could think as the composite, and the laboriously investigated as the simple. However, in scientific questions I want to deprive such feelings of authority: the contemporaries of Copernicus were equally directly conscious and felt that the earth did not revolve. Still, the most direct path would be to start from our immediate sensations and to show how by means of them we attained knowledge of the universe. However, since this does not seem to lead to our goal, let us follow the inverse path of natural science. We frame the hypothesis that complexes of atoms had developed that were able to multiply by forming similar ones round them. Of the larger masses so arising the most viable were those that could multiply by division, and those that had a tendency to move towards places where favourable conditions for life prevailed.

This was greatly furthered by the receptivity for external impressions, chemical constitution and the motion of the circumambient medium, light and shade and so on. Sensitivity led to the development of sensory nerves, mobility to motor nerves; sensations that through inheritance led to constant strong compelling messages to the central agency to escape from them we call pain. Quite rough signs for external objects were left behind in the individual, they developed into complicated signs for complex situations and, if required, even to quite rough genuine internal imitations of the external, just as the algebrist can use arbitrary letters for magnitudes but usually prefers to choose the first letters of the corresponding words. If there is such a developed memory sign for the individual himself, we define it as consciousness. In this there is a continuous path from the closely connected clear conscious ideas to those stored in memory and to unconscious reflex movements. Does not our feeling tell us once more that consciousness is something quite different? However, I have silenced feeling: if the hypothesis explains all the phenomena concerned, feeling will have to give way as in the question of the earth's rotation. It will be a much later question, but one that can be solved only in this way, how from our sensations which are the simplest elements of our thinking we were able to reach hypotheses. However, I must end, lest I become false to my resolution to leave metaphysics aside. Much of what I have said may be mistaken, but all of it reflects my conviction. Only if each continues to work where and how he is able, can we come nearer to truth; only, as Schiller says, by "busyness that never flags, that slowly forms but never destroys, that to the building of eternities brings grains of sand, yet from the heavy debt of time wipes minutes, days and years".

Thus I too shall be satisfied if my present lecture has contributed a grain of sand to the spread of the knowledge of nature.

NOTE

1 J. von Kries, *Die Prinzipien der Wahrscheinlichkeitsrechnung*, Freiburg i.B., 1886.

ON THE SIGNIFICANCE OF THEORIES*

When some days ago I learnt of the plan for today's ceremony, it was at first my firm intention to ask you to refrain. For how, I asked myself, can an individual deserve being honoured in this way? Surely, all of us are just collaborators in a great enterprise, and everyone who does his duty in his post deserves equal praise. If therefore an individual is singled out from the community this can in my view never be aimed at him as a person but only at the idea that he represents; only by completely giving himself over to an idea can the individual gain enhanced importance.

Therefore I decided not to insist on my request only when I related all honours not to my own modest self but to the idea that fills my thought and action: the development of theory, for whose glory no sacrifice is too great for me; since theory is the whole content of my life, let it likewise be the content of my present words of thanks.

I should not be a genuine theoretician if I were not first to ask: what is theory? The layman observes in the first place that theory is difficult to understand and surrounded with a tangle of formulae that to the uninitiated speak no language at all. However they are not its essence, the true theoretician uses them as sparingly as he can; what can be said in words he expresses in words, while it is precisely in books by practical men that formulae figure all too often as mere ornament.

A friend of mine has defined the practical man as one who understands nothing of theory and the theoretician as an enthusiast who understands nothing at all. The rather pointed view contained in this we will likewise oppose.

I am of the opinion that the task of theory consists in constructing a picture of the external world that exists purely internally and must be our guiding star in all thought and experiment; that is in completing, as it were, the thinking process and carrying out globally what on a small scale occurs within us whenever we form an idea.

* *Populäre Schriften*, Essay 5. In reply at a farewell ceremony, 16 July 1890, at Graz, when the author had been called to a professorship at Munich.

It is a peculiar drive of the human spirit to make itself such a picture and increasingly to adapt it to the external world. If therefore we may often have to use intricate formulae to represent a part of the picture that has become complicated, they nevertheless always remain inessential if most serviceable forms of expression, and in our sense Columbus, Robert Mayer and Faraday are genuine theoreticians. For their guiding star was not practical gain but the picture of nature within their intellect.

The immediate elaboration and constant perfection of this picture is then the chief task of theory. Imagination is always its cradle, and observant understanding its tutor. How childlike were the first theories of the universe, from Pythagoras and Plato until Hegel and Schelling. The imagination at that time was over-productive, the test by experiment was lacking. No wonder that these theories became the laughing stock of empiricists and practical men, and yet they already contained the seeds of all the great theories of later times: those of Copernicus, atomism, the mechanical theory of weightless media, Darwinism and so on.

In spite of all mockery the drive to form a theoretical view of external things remained unconquerable in the human breast and it constantly gave rise to new flowers. As Columbus set course always towards the west, so this drive always unswervingly directed us towards this great goal.

When in the end sober experimental understanding and the dexterity needed for handling the many invented devices and machines came increasingly into their own, the old and variegated imaginative structures were sifted and refined and, with amazing speed, gained in truth to nature and in importance. Today one can aver that theory has conquered the world.

Who can see without admiration how the eternal stars slavishly obey the laws that the human spirit has not indeed given to but learnt from them. And the more abstract the theoretical investigation, the more powerful it becomes. If, still somewhat mistrusting the path on which, being led by formulae rather than leading them, we have reached a theorem of arithmetic, we test it on numerical examples, we are even more strongly haunted by the feeling that numbers without exception must inevitably bow to our formulae.

But even those who value theory only as a milch cow, can no longer doubt its power. Are practical disciplines all of them by now not penetrated by theory and do they not all follow their reliable guiding star? The forms of Kepler and Laplace not only show the stars their celestial courses, but along with Gauss's and Thomson's calculations on the earth's magnetism they show ships their way on the high seas. The gigantic structures of the Brooklyn Bridge that stretches beyond sight and the Eiffel tower that soars without end rest not only on the solid framework of wrought iron, but on the solider one of elasticity theory. Theoretical chemists have become rich through practical application of their syntheses, not to mention the electrical engineer! Does he not pay constant homage to theory by the fact that next to pound and penny the names that are most familiar to him are Ohm, Ampère and so on, all of them great theoreticians, none of whom, alas, were blessed by the lucky fate of the chemists just mentioned; for their formulae did not become fruitful in practice until after they had died. Indeed, it may well not be long before these great electric theorists will be glorified in every domestic bill and in the next century every cook may know with how many 'Volt-Ampères' one fries meat and how many 'Ohms' her lamp has got.

It is precisely the practical technician who as a rule treats the complicated formulae of electric theory with a surer hand than many a tiro scientist, because he has to pay for his errors not only by way of reproof from his teacher but in hard cash. Indeed, almost any carpenter or metal worker knows today how much a grasp of descriptive geometry, the theory of machines and so on make him more competitive. I must mention also the splendid field of medical sciences, where theory gradually seems to gain a footing too.

One is almost tempted to assert that quite apart from its intellectual mission, theory is the most practical thing conceivable, the quintessence of practice as it were, since the precision of its conclusions cannot be reached by any routine of estimating or trial and error; although given the hidden ways of theory, this will hold only for those who walk them with complete confidence. A single mistake in a drawing can multiply a result a thousandfold, whilst an empirical worker never errs so far; for that reason there will no doubt always remain some cases where the thinker who is immersed in his ideas and always bent on what is general will be outdone by the clever and self-interested practical man; witness Archimedes who fell victim to the attacking Roman, or another Greek philosopher who, while looking at the stars, stumbled over a stone. Let silence overtake the question "what is the use of it?" which is customarily thrown at any more abstract endeavours. One would like to ask the counter-question: "what is the use of furthering life by gaining mere practical advantages at the expense of that which alone makes life worth living, namely the tendance of the ideal?"

However, theory keeps well away from overrating itself; its very defects are grounded in its own nature and it is theory itself that uncovers its own errors; indeed, already Socrates placed the main emphasis on the recognition of the gaps in his own knowledge. All our ideas are purely subjective. That this is so even as regards our views on being and not-being is shown by Buddhism which reveres nothingness as the really existing. I called theory a purely mental inner picture, and we saw to what a high degree of perfection this may be brought. How then, as we become more and more immersed in theory, could we fail to take the picture for what really exists? It is in this sense that Hegel is said to have regretted that nature was unable to realize his philosophic system in its full perfection.

Thus it may happen to the mathematician, who is constantly occupied with his formulae and blinded by their inner perfection, that he takes their mutual relations as the really existing and turns away from the real world. What the poet laments then holds of the mathematician, that his works are written with the blood of his heart and highest wisdom borders on supreme folly. It is in this sense, too, that I take Goethe's dictum about the greyness of theory compared with life, a saying that cannot be avoided when one discusses this subject; he was indeed a theoretician through and through according to our conception, although of course avoiding this aberration. Incidentally, he puts this sentiment in the mouth of the devil, who later says with a sneer: "Do but spurn reason and science... you will be unconditionally mine!"

If at the outset I have declared myself an advocate of theory, I will not deny that I have myself experienced the evil consequences of its spell. Yet what would be more effective against this spell, what could drag us back more forcefully into reality than the living contact with so honourable a gathering as this present one? For this kindness that you have shown me I thank you all: first you, Rector, who organised this ceremony, next, the orator, colleagues and guests who followed his call, and finally the gallant sons of our alma mater, whose strong endeavours and noble enthusiasm were my support through 18 years. May Graz university grow and flourish and always be and remain what is highest in my view: a stronghold of theory!

ON ENERGETICS*

A discussion such as this present one on energetics is not undertaken in the expectation that one side will be right and the other wrong, but with the intention that the views will be clarified. Therefore I can be satisfied with the result as regards the relations between energetics and mechanics. Helm's latest essay (*Wied. Ann.* **57**, 1896 p. 646) seems to put everything perfectly straight.

Planck and Helm have shown (simultaneously, as now turns out) that the ordinary equations of motion for a system of material points can be obtained from the principle of energy if we assume that it holds separately for each of the particles in the direction of every co-ordinate or, according to Helm, in any arbitrary direction whatsoever.

On the other hand Helm goes so far as obtaining the Lagrange equations and thus the whole of mechanics by transformation of rectangular coordinates of material points and of the forces acting on them, which therefore involves the presupposition that bodies are systems of material points. This presupposition, however, evidently once more takes us completely into the area of atomism. From it follows in known ways that for long-lasting motion under the influence of forces that do not act uniformly on all material points, there must arise irregular mutual motions of the particles,[1] which always swallow up a part of the visible kinetic energy; that if the motion is sufficiently violent, the particles creep past each other, which liquefies the body; and that particles must separate from the surface, which vaporizes the body.

These atomistic hypotheses incidentally recognize the concept of energy, too, as one of the most important; indeed, if you will, they might even be obtained from that concept by means of suitable subsidiary assumptions. If however energetics will not recognize such hypotheses on the ground that they are insufficiently attested, it would have to take quite a different path.

* *Populäre Schriften*, Essay 9. First published in *Annalen der Physik und Chemie* **58** (1896) 595.

Just how one might construct a mechanics on the assumption that the kinetic energy of the motion is the primarily given and the moving object itself a concept derived from it, I cannot quite imagine at present. If then energetics takes the comfortable path of starting from the concept of mass, then in order to avoid the atomistic hypothesis it would have to assume that matter continuously occupies its space. From the principle of energy together with suitable auxiliary hypotheses one would then first have to obtain the equations of motion for rigid bodies, perhaps by deriving Lagrange's equations without the detour via the co-ordinates of the individual points of which the body consists and via the forces that act on them. By means of fresh auxiliary hypotheses one would have to derive from the formulae for elastic and hydrodynamic energy the corresponding equations of motion. All these derivations should be possible, indeed variously so, according as this or that auxiliary hypothesis is enlisted, and I should regard it as useful to science to attempt such derivations.

What would seem to be more difficult is to give a survey purely from the point of view of energetics of all cases where mechanical energy is transformed into heat, phenomena of melting and vaporisation, properties of gases and vapours and so on, whereas it is precisely these phenomena that become so intelligible by means of molecular theory and the special mechanical theory of heat.

Energetics seems as yet a long way from having solved all the problems sketched here. It is clear that until this has happened no judgment of how intuitive the auxiliary hypotheses needed by energetics are can be formed, nor can they be compared with molecular theory over the whole range of mechanics.

The thermodynamic equation that I originally criticised has now also been given a clear meaning by Helm, since he affirms that J is here not the internal intensity within the body but the intensity of the external reaction, which makes the proposition clear and intelligible at least when J stands for pressure. However, I think that in this connection many of Helm's other explanations must be made more precise; for wherever he applies the proposition in question (*Mathematische Chemie*, pp. 45, 46, 47, 60) it seems as though in contradiction with the present definition he slides back into taking J as the internal intensity so that again he would have to write the equal sign only.

However, this point is naturally rather inessential, and only if a clear and unobjectionable account of thermodynamics, chemistry and electricity from an energetic point of view had been achieved at least in its first basic outlines, would it be possible to ascertain what essentially new additions energetics has made to Gibbs's theory.

During proof-correction I come across H. Ostwald's reply (p. 154). This seems to show that, contrary to my previous view, he does not regard energy as the originally given in mechanics and proposes to deduce mass from certain properties of it, but that he retains the concepts of the old mechanics, starting from mass and defining energy as $\frac{1}{2}mv^2$. Whether after that one speaks of mass or energy or both as the substantially existing, or perhaps of neither but of our ideas instead, all this in view of the retention of the old ideas, would seem to be hardly more important than whether the masses or energy units are used as base for the system of measurement. As regards the rest of the reply, I think I can be brief.

That H. Ostwald is personally convinced of his approach and will not let himself be shifted from it I have never doubted. In research, impulses that are not clearly conscious obviously defy discussion. However, as regards the alleged barrenness of atomism, many chemist too will disagree, since they are wont to deduce the possible number of isomeric compounds and the property of rotating the plane of polarisation directly from the picture that they have formed of the position of the atoms. For my part I permit myself to point out that in justifying his theorems Gibbs must surely have used molecular ideas, even if he nowhere introduced molecules into the calculation; that the theorems on energy and entropy of gases, of dilute solutions and above all on those of a mixture of a dissociating body with its constituents were discovered and justified only through the conception that the various molecules exist adjacently in space; finally, that the most recent electro-chemical theory has its starting point in the purely molecular view that Nernst had of the pressure of solutions. It was only later that these propositions were severed from their molecular justification and presented as pure facts. The mathematical part of gas theory on the other hand pursues mainly the purpose of further development of mathematical method, for the valuation of which immediate practical utility was never decisive. Let the purely practical man skip this part but also forebear to criticize.

NOTE

1 Even assuming space continuously occupied, does it not follow from the equations of elasticity that analogous irregular vibrations of the volume elements must arise, thus offering the most obvious explanation of transformation of elastic vibrations into heat?

ON THE INDISPENSABILITY OF
ATOMISM IN NATURAL SCIENCE*

Beside atomism in its current form a second method is customary in theoretical physics, namely that of representing by means of differential equations as strictly circumscribed an area of facts as possible. We will call it phenomenology on a mathematico-physical basis. Since this gives a new picture of the facts and since it is of course advantageous to have as many pictures as possible, it is naturally of great value alongside atomism in its present form. Another phenomenology, which I will call the energetic kind, is to be mentioned later. Now it has often been said that the pictures obtained from the phenomenological method deserve preference over atomistic ones, and that for internal reasons.

I tend to shun such general philosophical questions, so long as they have no practical consequences, for they cannot be framed as precisely as special questions so that answering them is more a matter of taste. However, it seems to me as if at present atomism, for the hardly valid reason just mentioned, is being neglected in practice and therefore I thought I should do my bit to prevent the damage that in my opinion might accrue to science if phenomenology were now to be raised to the status of dogma, as atomism was previously.

To avoid misunderstandings, I will from the outset denote the purpose of the following considerations as being the answering of certain very specific questions. Since the profit that atomism in its development has given to science will not be doubted by any impartial expert in the history of science, we may formulate the question thus: has not atomism in its present form also great advantages over the current form of phenomenology? Is there any likelihood that in the foreseeable future phenomenology could develop into a theory that possesses those same advantages so peculiar to atomism? Alongside the possibility that current atomism may one day be abandoned, is there not also another that phenomenology

* *Populäre Schriften*, Essay 10. First published in *Annalen der Physik und Chemie* **60** (1897) 231.

will more and more dissolve in it? Finally would it not be to the detriment of science if one were not to go on cultivating current atomist views as assiduously as phenomenological ones even today? The answer to these questions, let me say at once, will be favourable to atomism, as a result of the considerations that follow.

The differential equations of mathematico-physical phenomenology are evidently nothing but rules for forming and combining numbers and geometrical concepts, and these in turn are nothing but mental pictures from which appearances can be predicted.[1] Exactly the same holds for the conceptions of atomism, so that in this respect I cannot discern the least difference. In any case it seems to me that of a comprehensive area of fact we can never have a direct description but always only a mental picture. Therefore we must not say, with Ostwald, "do not form a picture", but merely "include in it as few arbitrary elements as possible".

Mathematico-physical phenomenology sometimes combines giving preference to the equations with a certain disdain for atomism. Now in my view, the assertion that a differential equation goes less beyond the facts than the most general form of atomistic views rests on a circular argument. If from the outset you hold that our perceptions are represented by the picture of a continuum, then indeed differential equations do not, while atomism does, go beyond this presupposition. Quite otherwise if one is used to thinking atomistically, then the position is reversed and the conception of the continuum seems to go beyond the facts.

Let us for example analyse the meaning of the classical instance of Fourier's equation for heat conduction. It expresses nothing else but a rule consisting of two parts:

(1) Within a body (or, more generally, in a regular arrangement within a corresponding bounded three-dimensional manifold), imagine numerous small things (let us call them elementary particles or, better still, elements or atoms in the most general sense), each of which has an arbitrary initial temperature. After a very short time has elapsed (or when the fourth variable increases by a small amount) let the temperature of each particle be the arithmetic mean of the initial temperatures of the particles that had immediately surrounded it previously.[2] After a second equal lapse of time the process is repeated and so on.

(2) Imagine both the elementary particles and the increments of time becoming ever smaller and their number growing in corresponding

proportion, and let them stop at those temperatures at which further diminution no longer noticeably affects the results.

Likewise, definite integrals that represent the solution of the differential equation can in general be calculated only by mechanical quadratures and thus again demand division into a finite number of parts.

Do not imagine that by means of the word continuum or the writing down of a differential equation, you have acquired a clear concept of the continuum. On closer scrutiny the differential equation is merely the expression for the fact that one must first imagine a finite number; this is the first prerequisite, only then is the number to grow until its further growth has no further influence. What is the use of concealing the requirement of imagining a large number of individuals now, when at the stage of explaining the differential equation one has used that requirement to define the value expressed by that number? My apologies for the somewhat banal expression, if I say that those who imagine they have got rid of atomism by means of differential equations fail to see the wood for the trees. Explaining differential equations by complicated geometrical or other physical concepts would indeed help all the more to make the equation for heat conduction appear in the light of an analogy rather than of a direct description. In reality we cannot distinguish the neighbouring parts. However, a picture in which from the start we did not distinguish adjacent parts would be hazy; we could not apply to it the prescribed arithmetical operations.

If then I declare differential equations, or a formula containing definite integrals, to be the most appropriate picture, I surrender to an illusion if I imagine that I have thus banished atomistic conceptions from my mental pictures. Without them the concept of a limit is senseless; I merely add the further assertion that however much our means of observation might be refined, differences between facts and limiting values will never be observable.

Does not therefore the picture that presupposes a very large but finite number of elementary particles go less beyond the facts? Has the position not been reversed? Whereas in the past the assumption of a definite size of atoms was regarded as a rough conception going arbitrarily beyond the facts, now it seems to be precisely the more natural one, and the assertion that differences between facts and limiting values can never be discovered because till now (perhaps not even in all cases) they have not yet been discovered, adds something new and unproved to the picture. Why this assertion, patched on after the event, should make the picture clearer, simpler or more likely I cannot grasp.[3] Atomism seems inseparable from the concept of the continuum. The reason why Laplace, Poisson, Cauchy and others started from atomistic considerations is evidently that in those days scientists were as yet more clearly conscious that differential equations are merely symbols for atomistic conceptions so that they felt a stronger need to make the latter simple. The first forms of atomism we might compare with the complicated verbiage that ancient physicists indulged in rather than calculate with named quantities, while getting used to the symbols of integral calculus resembles getting used to expressions like cm s^{-1}. The convenience thus achieved may however lead to many faulty inferences if one forgets the meaning arbitrarily given to division by a second.

As with the equation for heat conduction, the basic equations of elasticity can be generally solved only if one first imagines a finite number of elementary particles that act on each other according to certain simple laws and then once again looks for the limit as this number increases. This limit is thus once again the real definition of the basic equations and the picture that from the outset assumes a large but finite number seems once more simpler.

In this way, by attributing to the atoms in question only such properties as are needed to describe a small factual domain as simply as possible, we can obtain for each such domain a special atomism,[4] which it would seem is no more a direct description than what is ordinarily called atomism but at least constitutes a picture reasonably free from arbitrary features.

Now phenomenology tries to combine all these special atomisms without prior simplification, in order to represent actual facts, that is in order to adapt all conceptions contained in these atomisms to the facts; however, since these are countless concepts severally taken from small factual domains and hardly compatible, along with countless differential equations each with its own peculiarities in spite of many analogies, we must from the outset expect that the representation will turn out very complicated. Indeed we find that if phenomenology is to do no more than represent the interlinkage between a few domains of phenomena of quasistatic processes (elastic deformation with heating and magnetisation and the like), one needs already very unwieldy and enormously complicated equations. Besides even then one has to introduce hypotheses and thus go beyond the facts (for example if one wants to represent the dissociation of gases after Gibbs and that of electrolytes after Planck).

To this we must add the circumstance that all concepts of phenomenology are derived from quasi-stationary processes and no longer hold good for turbulent motion. For example, we can define the temperature of a body at rest by means of a thermometer inserted in it. If the body moves as a whole, the thermometer can move with it, but if every volume element of the body has a different motion the definition becomes void and it is likely or at least possible that the different energy forms can no longer be sharply separated (what is heat and what is visible motion and so on).

Considering this and the complication taken on by phenomenological equations even in the few cases where the interlinkage of several domains of phenomena has been represented, one may form some idea of how difficult it is to use this method to describe arbitrarily turbulent phenomena perhaps involving chemical reactions; that is, without prior adjustment to each other by means of simplifications that are of course arbitrary of the atomisms corresponding to the different factual domains. Compared with the properties that one would thus have to ascribe to the elementary particles, Lemery molecules would be veritable paradigms of simplicity.

One special phenomenology, which I will call energetic (in the widest sense), hopes to bring the various atomisms corresponding to individual phenomenal domains closer together, by further pursuing what is common to all domains. Two kinds of such common features are known. To the first belong certain general propositions such as the principles of energy, entropy and so on, what we might call general integral propositions valid in all domains. The second consists in analogies that can pervade the most varied domains, and are often based only on identity of form which certain equations must assume when certain approximations are made while the analogies often seem to cease as regards finer details. (Approximate proportionality of small changes of a function with those of its argument, remainder of the first or second differential quotient with roughly constant coefficients, linearity of small quantities and hence superposition. Analogies in the behaviour of different energy forms, too seem partly to rest on such purely algebraic reasons.) Yet in spite of the enormous importance

of the integral propositions (because of their universal validity and consequent high certainty) and the analogies (because of the many computational advantages and new perspectives that they offer), they never
furnish more than a small part of the total complex of facts; to represent
each individual domain of phenomena more precisely thus already required so many special additional pictures (natural history of the domain
in question) that, as I think I have amply shown elsewhere, so far nobody
has succeeded in giving even an unambiguous and comprehensive description of a single domain of stationary phenomena by means of this method,
let alone a survey of all phenomena including turbulent ones. The question whether this path will one day lead to comprehensive pictures of
nature is thus for the time being purely academic.

To get closer to this last goal, current atomism does indeed seek to
reconcile the foundations of the various phenomenological atomisms by
arbitrarily completing and altering the properties of the atoms required
for the various factual domains in such a way that they may serve for
representing many domains at the same time.[5] In a manner of speaking,
atomism resolves the properties of the atoms required for the individual
factual domains into components (see note 4 above) in such a way that
the latter fit several domains. This is obviously not possible without a
certain arbitrariness that goes beyond the facts, just as the resolving of
forces into components does.[6] However, it obtains the compensating
advantage of being able to give a simple and perspicuous picture of a far
greater sum of facts.

While phenomenology requires separate and mutually rather uncon-
nected pictures even for the mechanical motion of centres of gravity and
rigid bodies, for elasticity, hydrodynamics and so on, present day atomism
is a perfectly apt picture of all mechanical phenomena, and given the
closed nature of this domain we can hardly expect it to throw up further
phenomena that would fail to fit into that framework. Indeed, the picture
includes thermal phenomena: that this is not so readily proved is due
merely to the difficulty of computing molecular motions. At all events all
essential facts are found in the features of our picture. Further, it proved
itself extremely useful for representing crystallographic facts, the constancy of proportions of mass in chemical compounds,[7] chemical isomer-
isms, the relations between the rotation of the plane of polarisation and
chemical constitution, and so on.

For the rest, atomism remains capable of being developed much further. One may conceive of atoms as more complicated individuals endowed with arbitrary properties, as for example the vector atoms which, as we saw in note 4 above, at present furnish the simplest description of electro-magnetic phenomena.[8]

As regards turbulent phenomena, so far quite inaccessible to phenomenology, current atomism does of course approach them with definite presuppositions; however, it possesses valuable hints on how these phenomena might be represented and in some cases can positively predict them. Thus gas theory can predict the course of all mechanical and thermal phenomena in gases even under turbulent motion and therefore gives indications on how, for these phenomena, one will have to define temperature, pressure and so on. It is precisely the main task of science to fashion the pictures that serve to represent a range of facts in such a way that we can predict from them the course of other similar facts. Naturally, it is understood that the prediction must still be tested by experiment. Probably it will be verified only in part. There is then some hope that one might modify and perfect the pictures in such a way that they do justice to the new facts too. (We learn something new about the constitution of atoms.)

One may of course justifiably demand that the picture must not be given additional arbitrary features beyond what is absolutely necessary for the description of wider areas of phenomena (since arbitrariness must be confined to the most general level possible) and that people should always be ready to modify the picture or even keep in mind the possibility of recognizing that a given picture had better be replaced by quite a new and basically different one. The fact that the construction of the new picture would have to be based on the so far untouched special phenomenological pictures is reason enough to cultivate these with care as well, alongside atomism.

Finally, I should like to go further and almost venture the assertion that it lies in the nature of a picture to have to add certain arbitrary features for the purpose of representation, and that strictly speaking one goes beyond experience as soon as one infers from a picture adapted to certain facts to even a single new fact. Is it mathematically certain that in order to represent all facts one will not have to replace Fourier's equation for heat conduction by quite a different relation which reduces to the former only in cases hitherto observed, so that for any arbitrary new observation we should have to alter the picture totally and therefore also our conceptions as regards heat exchange amongst the smallest parts? For example, all bodies examined in the past might happen to exhibit certain regularities without which Fourier's equation becomes false.

Just like Fourier with the law of specific heat and the fact that heat exchange between two touching bodies is proportional to the temperature difference between them, so gas theory with the general laws of mechanics and the fact that bodies displace each other on contact but no longer affect each other at somewhat greater distances: both carry them over to the smallest parts, which as we saw are indispensable if one is to represent extended bodies. The assumption that one and the same kind of smallest parts suffices for representing the liquid and gaseous state of aggregation seems to me likewise well-founded, since the two states are continuous and this is the only hypothesis that answers to the demand for simplicity in describing nature. Admitting that these last two assumptions are justified we cannot, however, escape the consequence that the smallest parts are set into invisible relative motion that swallows up visible kinetic energy while it is surely not unlikely that this motion will be perceived by certain nerves (the special mechanical theory of heat), and that in very dilute bodies the particles mostly travel in nearly straight lines (kinetic gas theory). The picture by which we represent mechanical phenomena would merely become more complicated if not contradictory, were we to omit these inferences. The further assumption that the molecular motions never cease while provoked and visible motions gradually go over into molecular ones, is likewise quite in conformity with recognized mechanical laws.

All inferences from the special mechanical theory of heat, however disparate the fields in which they belong, have been confirmed by experience, indeed I would say that right down to the finest details they are peculiarly in tune with the heart-beat of nature.[9]

Of course Fourier's assumptions about heat conduction are so extraordinarily simple and the further facts that might still be computed from them so conformable to those already tested by observation, that it may perhaps seem to be splitting hairs to assert that Fourier's assumption and his equation (as a first approximation) are not absolutely certain. However, I do not find it strange that rather simple and plausible assumptions will do as soon as the factual domain is thus arbitrarily restricted, and that cases significantly different from those already tested will soon give out.

Should it ever become possible to construct as comprehensive a theory as current atomism on as clear and unobjectionable a basis as Fourier's theory of heat conduction, this would of course be ideal. Whether this might be realized by subsequent unification of the initially unsimplified phenomenological equation or rather through the fact that continual adaptation and practical confirmation of current atomist views will in the end come asymptotically close to the evidence of Fourier's theory, that question is I think as yet quite undecided.[10] For even if the observations already available are held to be inconclusive, and they seem to include molecular motion in liquids and gases directly observed, we cannot deny the possibility of future conclusive observations (that is, such as will raise the probability to as high a level as desired). It therefore seems to me quite wrong to assert with certainty that pictures like the special mechanical theory of heat or the atomic theory of chemical processes and crystallization must vanish from science one day. One can ask only what would be more disadvantageous to science: the excessive haste implicit in the cultivation of such pictures or the excessive caution that bids us abstain from them.

It is well known how much the conceptions of atomism have benefited physics, chemistry and crystallography, by making them more intuitive and perspicuous. We shall not deny that, especially at a time when these conceptions were as yet much less adapted to phenomena and were viewed from a more philosophic angle, they could be a hindrance as well and therefore in some cases appear like useless ballast. Nothing will be lost in certainty, while perspicuity will be retained, if we strictly separate the phenomenology of the best-attested results from atomistic hypotheses that serve comprehensiveness, both being further developed with equal vigour as being equally indispensable, rather than assert with one-sided regard for the advantages of phenomenology that one day it will certainly displace current atomism.

Even if it is possible to unite phenomenological pictures into a comprehensive theory along lines different from those of today's atomism, the following is certain:

(1) This theory cannot be an inventory in the sense that every single fact is denoted by a special sign; that would make it just as unwieldy to

find one's way about as actually to live through all these facts. Thus, like current atomism, it can be only a directive to build oneself a picture of the world.

(2) If one is not to harbour illusions as to the meaning of a differential equation or indeed of any continuously extended magnitude, such a picture must beyond doubt be essentially atomistic, that is an instruction to imagine, according to definite rules, the temporal changes of a very large number of things arranged in a manifold of presumably three dimensions. These things can of course be the same or different in kind, invariable or variable. This picture might correctly represent all phenomena if we assume the number to be large but finite, or maybe in the limit when the number grows indefinitely.

Imagine there could be an all-encompassing picture of the world in which every feature has the evidence of Fourier's theory of heat conduction, then it remains so far undecided whether we should reach that picture more readily by the phenomenological method or by constant further development and experimental verification of the pictures of current atomism. One might then equally well imagine that there could be several world pictures all of which possessed the same ideal property.

Note A. From the principles of this essay it follows no doubt that continuous geometrical figures such as the circle signify merely that we must first think of it as consisting of a finite number of points which must then be allowed to grow indefinitely. The limit approached by the perimeter of the inscribed and circumscribed polygon of n sides as n increases is precisely the definition of π. Yet the circle as geometrical concept will not be conceived as formed by a finitely large number of atoms, since it is not a thought-symbol for an individual constant complex, like the concept of one gram of water at $4\,°C$ and atmospheric pressure, but, like the concept of number, is to be applicable to the most varied complexes with the most varied (always very large) numbers of atoms.

Note B. What at the beginning we called 'elementary bodies' or 'atoms in the widest sense' or 'elements' can of course be given any other name, for example 'units of conception' or 'somethings'. However I would advise against the term 'volume-elements'. Firstly, it carries many conceptions that are precisely to be avoided to keep the picture clear, for example the idea of a definite shape (perhaps parallelepiped) or the idea that every element consists of smaller ones still that have the same property again in different degree (in the case of heat conduction, different temperatures). However this is precisely the most confused assumption and can never be made in the mechanical computation of definite integrals or of definite values defined by differential equations: there can be no heat conduction within the elements themselves. Secondly, the concept of 'volume element' is too narrow in other respects. How, for example, could we call vector atoms 'volume-elements'?

NOTES

[1] Cf. Mach, *Prinzipien der Wärmelehre*, Leipzig 1893, p. 363. His writings on these matters have greatly helped in clarifying my own world view.

[2] Maxwell, *Treatise on Electricity*, 1873, Vol. 1, sec 29; Mach, *loc. cit.* p. 118.

[3] Visual perceptions correspond to the excitation of a finite number of nerve fibres and are thus probably better represented by a mosaic than by a continuous surface. Similarly for the other senses. Is it then not likely that models for complexes of perceptions had better be composed of discrete parts?

[4] If we are honest, Hertz's assertion that his theory of electromagnetic phenomena consists in a certain system of differential equations can be made to signify only that he pictures these phenomena to himself by means of two kinds of conceptual objects, tightly filling space and both vectorial in character, and their change with time as to intensity and direction, dependent only on the immediate neighbourhood, as in heat conduction, though in a somewhat more complicated if readily specifiable manner. This amounts to an atomistic theory of electro-magnetism with the minimum of arbitrary elements. The demand for mechanical explanation of electromagnetism coincides with the desire to remove the complexity of that picture and its incongruence with pictures used in other areas, an incongruence that remains of course unnoticed if one merely compares the aspect of the differential equations. Evidently it is this incongruence and the likelihood of there being simpler pictures that is being expressed when one says one does not know what electricity is. Current phenomenology has thus quite returned to the position of Lemery (Ostwald, *Lehrbuch der allgemeinen Chemie* 2.II, 2nd edn., pp 5, 103), who likewise did not hesitate to ascribe the most complicated properties to atoms, so soon as this offered some explanation of the facts known to him; only we fail to notice it because we hide our heads in differential equations as an ostrich in sand.

If one recalls the meaning of the concept of limit, the ordinary equations of elasticity, as soon as they contain the displacements u, v, w and the elastic forces X_x, X_y ..., represent fairly complicated rules for the change of co-ordinates $x+u$, $y+v$, $z+w$ of ordinary points and the simultaneous change of vector atoms. Even the equations obtained when elastic forces are eliminated require further reductions before yielding the usual atomistic picture of elastic phenomena. In order to obtain that picture one has thus carried compositions or resolutions on the equations or on the pictures identical with the latter, just as in mechanics one composes or decomposes forces in order to obtain a suitably simple description.

Differential coefficients with respect to time likewise require that in our picture of nature we begin by taking time as divided into very small finite parts or atoms of time. If therefore we drop as not so far proved by experience the notion that there could be no discoverable deviation from the limit to which the picture approaches for ever diminishing atoms of time, we should have to imagine that even the laws of the mechanics of material points were only approximately correct. Just to give an idea what varied pictures might be chosen let me mention a special one here. Imagine a great number of spheres in contact in space (or rather in a three-dimensional manifold). According to a law A, to be discovered, their arrangement changes from one time atom to the next by a very small but finite amount. The variously shaped gaps between the spheres take the place of atoms in the old picture, the law A is to be chosen so that the change with time of the gaps affords a picture of the world. If it were possible to find such a picture that showed more comprehensive agreement than ordinary atomism, the picture would

52   FROM 'POPULÄRE SCHRIFTEN'

thereby be justified. Thus the view of atoms as material points and of forces as functions of their distance is no doubt provisional but must at present be retained failing a better one.

Of course, elementary reflection and experience alike will tell us that it would be hopelessly difficult to hit at once upon appropriate world pictures merely by aimless guesswork; on the contrary, they always emerge only slowly from adaptation of a few lucky ideas. Rightly, therefore, epistemology is against the doings of those framers of hypotheses who hope to find without effort a hypothesis that would explain the whole of nature, as well as against metaphysical and dogmatic foundations of atomism.

[5] The above account is of course not asserting that phenomenological equations have always temporally preceded the progress of current atomism. Rather, most of these equations were themselves obtained by considerations concerning specialized atoms taken from a different area of mechanical phenomena and did not acquire their phenomenological character until they had later been severed from those considerations. This is hardly surprising, since we have recognized that what these equations really signify is a demand for atomistic pictures, indeed it only reinforces the case of atomism.

[6] Such a feature arbitrarily ascribed to the picture of atoms is their invariability. The objection, that this was an unjustified generalisation of invariability of solid bodies observed for only a finite time span, would certainly be justified as soon as one tried to prove atomic invariability on a priori grounds as used to be the custom. However, we merely take this feature into the picture in order that the latter should be able to represent the essential concept of the greatest number of individual phenomena, just as one takes the first time derivative and the second space derivatives into the equation of thermal conduction in order that it should fit the facts. We are prepared to drop invariability in those cases where some other assumption would represent the facts better. For example, the vector atoms of the aether, mentioned in Note (4) above, would not be invariable with time.

Thus, atomic invariability belongs to those notions that show themselves very serviceable although the metaphysical considerations that led to it will not stand up to unprejudiced criticism. However, just because of this many-sided usefulness one must allow a certain likelihood that so-called radiant energy may be represented by pictures similar to those for matter (that is, that the luminous aether is a substance).

[7] No chemical reaction occurs instantaneously, but it is propagated in space with finite if large velocity. If therefore one applies the above analysis of the concept of continuity, the Mach-Ostwald picture of chemism (Mach, *loc. cit.*, p. 359) would state that elementary particles a, b respectively of the two substances vanish and in their place particles c of a new substance supervene. The difference between this and the customary views of chemical action is clearly no longer important. Nothing in this would be altered if it was only the limit, to be obtained by well-known procedures, that represented the facts.

[8] If by a mechanical explanation of nature we understand one that rests on the laws of current mechanics, we must declare it as quite uncertain whether the atomism of the future will be a mechanical explanation of nature. Only insofar as it will always have to state the simplest possible laws for temporal change of many individual objects in a manifold of probably three dimensions, can it be called a mechanical theory, at least in a metaphorical sense. If it should for example turn out to be impossible to find a simpler description of electromagnetic phenomena, one would have to retain the vector atoms discussed in the text above. Whether the laws according to which

these change with time are to be called mechanical or not will be entirely a matter of taste.

[9] Amongst many things I here mention only the explanation of the three states of aggregation and their transitions into each other, and the agreement of the concept of entropy with the mathematical expression of the probability or disorder of a motion. The assertion that a system of very many bodies in motion tends, bar unobservably few exceptions, to a state for which a specifiable mathematical expression denoting its probability becomes a maximum does seem to me to say more than the almost tautological statement that the system tends towards the most stable state. By the way, Mach (*loc. cit.*, p. 381) rightly surmises that when preparing a popular lecture on this subject I did not know the writings on the tendency towards stability which he quotes; indeed all but one of them appeared many years after my lecture and all of them after publication of those papers of which my lecture merely gives a popular version.

If the principle of energy were the only basis for the special theory of heat and the explanation of the principle the only purpose of the theory, then the latter would be superfluous, given the universal recognition of the former. However, we saw that many other reasons support the theory and that it affords a picture for many other phenomena as well.

The theory of electric fluids was from the start unnatural in quite a different way and was always recognized by many scientists as provisional.

[10] Important developments and further adaptations (cf. Mach, *loc. cit.*, p. 380) will however be necessary for both theories. Fourier's equation for heat conduction $du/dt = k\Delta u$ is definitely false for constant k. That with variable k it should have to take the form $h\, du/dt = (d/dx)(k\, du/dx) + (d/dy)(k\, du/dy) + (d/dz)(k\, du/dz)$ is surely not sufficiently confirmed by experience. It does not represent the reaction of the compressions and dilatations, inevitably associated with non-stationary heat conduction, on the heat distribution, nor the direct action of hot volume elements on other distant ones by radiation in a diathermanous body (and who knows whether all bodies might not be diathermanous for certain rays that of course transmit energy and therefore heat as well). It is of course said that these effects do not belong to pure heat conduction; but such a pure phenomenon would once more be a metaphysical and hypostasized concept.

[*Editor's note:* For the notation, see my note on p. 12 above.]

MORE ON ATOMISM*

To make my complete agreement with all that Volkmann (*Wied. Ann.* **61** (1897) 196) says about this subject more sharp and precise, I wish to confirm that in my first essay on this subject [see p. 41] I never called into question the practical utility of the concept of elements of volume that in turn consist of smaller such elements and so on indefinitely; I merely declared this concept to be epistemologically inferior to atomistic conceptions. I will try once more to illustrate the idea I had in mind by means of the simple example of Fourier's equation for heat conduction.

Let us request somebody to imagine a large finite number of points in a given bounded space, perhaps the corners of a set of regularly stacked cubes. We require further that at the initial time he should associate a definite given temperature with each point. After a short finite while, let the temperature of each point become the arithmetic mean of the temperatures at the six neighbouring points. After a second equal time, a third temperature distribution is to be determined according to the same rule as the second, and so on. Thus we have given the person a quite definite instruction for definite computations, which under the prevailing regularities he may perhaps succeed in simplifying but will in any case be able to complete unambiguously and with certainty provided he is patient enough. Let us call this instruction a conceptual picture that is epistemologically unobjectionable, because it is clear and unambiguous, atomistic in the widest sense, because it is based on a finite number of elements.

We can now ask the person to imagine the points and time spans to be a milliard times more crowded, and then the same again and so on, until a further increase no longer has any noticeable influence on the temperature calculated for any point at any time. We then still have a picture that is atomistic and epistemologically unobjectionable. For whether the person can find the time to carry out all this calculation, whether he can

* *Populäre Schriften*, Essay 11. First published in *Annalen der Physik und Chemie* **61** (1897) 790.

shorten the determination of the relevant limit by means of computational tricks, that is his own affair. The task is clearly defined and solvable if enough time is spent.

In this I have deliberately left quite on one side the question whether we continue to get closer to the laws of thermal equilibrium in a real body the more we increase the number of elements, or whether the finest details of the heat exchange can be better represented by not increasing this number beyond a certain level and attributing to them more special properties, for example that they mutually radiate heat or convey it in the form of oscillatory motions. For I think it would be idle to speak about these latter questions, so long as nobody has succeeded in actually observing phenomena that point to the utility of such special pictures.

If however we tell somebody to imagine a body as consisting of elements of volume that can be infinitely divided into further such, the temperature at any point being a continuous function of the coordinates that satisfies some partial differential equation, this does not give him a usable rule that he could really apply to anything unless we explain the meaning of the partial differential equation starting with a finite number of elements as above.

The following would be a very simple analogy: if I tell somebody to sum the series $1 + \frac{1}{2} + \frac{1}{4} + \frac{1}{8} + \dots$ really to the extent of infinitely many terms, he will be unable to do it; but if I tell him to sum so many terms that a further increase will no longer noticeably influence the result, I have given him a clear and executable prescription, and all proofs that the sum of infinitely many terms equals 2 merely signify that if you add countless thousands of further terms you will never exceed 2, though you will approach it more and more.

The same, it seems to me, holds of the equations of elasticity theory and of the most complicated equations of mathematical physics.[1] When carrying out the by now customary manipulations with the symbols of integral calculus, one may temporarily forget that in forming these concepts we based ourselves on starting with a finite number of elements, but we cannot really circumvent this assumption.

That, too, seems to be the reason why groups of mutually interacting atoms of an elastic body are intuitively much clearer than interacting volume elements.

This naturally does not exclude that, once we have become used to the abstraction of volume elements and other symbols of integral calculus and have practised the methods of operating with them, it might be convenient and expedient no longer to remember the peculiarly atomistic meaning of these abstractions when we derive certain formulae that Volkmann calls those for coarser phenomena. These abstractions constitute a general schema for all cases where we may imagine the number of elements in a cubic millimetre to be 10^{10} or $10^{10^{10}}$ or milliards of times more still; hence they are indispensable especially in geometry, which must of course be equally applicable to the most varied physical cases where the number of elements can be very different. In using any such schemata it is often expedient to leave aside the basic idea from which they have sprung or even forget it for a while; but I think it would nevertheless be erroneous to believe that one had thus got rid of it.

For example, algebraic magnitudes are only general schemata for numerical values. In many calculations it would cause quite pointless delays always to substitute definite numerical values; indeed, we might very well divide calculations into two classes, those where we must descend to substituting numerical values and those where it is unnecessary, indeed superfluous and harmful. From an epistemological point of view however, the algebraic expressions in the latter kind of calculations are nonetheless nothing else but symbols for numerical values.

NOTE

[1] This is obviously not to say that the former equations could be explained only by means of the atoms of Navier and Poisson. Perhaps they could equally well be represented by quite different mental pictures which however, if epistemologically clear and unobjectionable, must once more be what we have called atomistic in the wider sense of the term.

The concepts of differential and integral calculus divorced from any atomist notions are typically metaphysical, if following an apposite definition of Mach we mean by this the kind of notion of which we have forgotten how we obtained it.

Obviously all properties of bodies that do not arise simply from the joint action of the large number of elements must be ascribed to the elements themselves; there is no other way of obtaining a picture of extended and apparently continuous bodies having these properties. For this reason I could never understand how it could be a reproach to atomism that it ascribed the properties of extended bodies to their elements as well.

ON THE QUESTION OF
THE OBJECTIVE EXISTENCE OF PROCESSES
IN INANIMATE NATURE*

Let me begin by defining my position by way of a true story. While I was still at high school my brother (long since dead) tried often and in vain to convince me how absurd was my ideal of a philosophy that clearly defined each concept at the time of introducing it. At last he succeeded as follows: during a lesson a certain philosophic work (I think by Hume) had been highly recommended to us for its outstanding consistency. At once I asked for it at the library, where my brother accompanied me. The book was available only in the original English. I was taken aback, since I knew not a word of English, but my brother objected at once: "if the work fulfils your expectations, the language surely cannot matter, for then each word must in any case be clearly defined before it is used."

It would be hard to show more drastically what wealth of experience, as well as of words and ideas used for denoting it, must be presumed as known if we are to understand each other at all, and that we cannot define everything but merely need use known signs to indicate rules for simplifying our ways of denoting and adapting them to experience.[1] Just as in geometry Euclid begins with unprovable axioms, we shall begin by examining what facts constitute the basis and precondition for knowledge. We shall honestly admit that with these facts we cannot and should not do more than recall them to memory by known signs, nor shall we be amazed that it is precisely these facts that have till now been regarded as the most difficult to explain.

Everyone knows what is meant by perceptions of the senses and impulses of the will. It is a precondition of intelligence that there are constant regularities between these,[2] which we can encompass by means of relatively few ideas. What this means is known by experience and we shall not find it puzzling that it can be as little further explained as the reason why these regularities occur. If moreover the sense perception (or perceptual complex) A following on the impulse of the will (or impulsive complex) B

* *Populäre Schriften*, Essay 12. First published in *Wien. Ber.* **106** (1897) Part IIa, 83.

always leads to a sense perception C, but following on impulse D to another sense perception E, this must leave within us certain impressions (memories, pictures of the world) which are of course related to actual processes as signs are to the things they denote (we say that after A and B we expect perception C, but after A and D, E); in many cases, these impressions must have the consequence that sense perception A will always be followed by impulse B but not D, and that the more surely the more formed the impressions. (We react to the impressions, they engage our feelings.) In that case we call C a desired perception and E an undesired one.[3] These impulses therefore depend in special ways on our inner states (memories). Hence we say that they proceed from us and call them voluntary, which of course is not to say that they obey no laws.[4]

Since through good memory images we attain things we desire, such images are themselves desired. It now turns out that by means of certain volitions we attain and refresh memories and even can complete and perfect their conjunction. Since we desire good memory pictures such impulses will often supervene (we imagine, reflect).

Actions that are followed by things we desire and ideas under whose guidance we act in this manner we denote as correct. We must aim at having ideas that are correct and economical as well, that is we are to be able always to reach the correct mode of action with the least expenditure of time and effort. The demand on any theory is that it be correct and economical; for on that very account it will then correspond to the laws of thought. I do not think that this needs to be set up as a special requirement, as Hertz has done.

The process described at the beginning may of course be extremely complicated. Suppose different perceptual complexes A_1, A_2, A_3, ... having certain common parts T (similarity) were always followed by a sensation C, or a volition B following it had provoked C. The impression that this leaves in our memory is marked thus: we expect that after each perceptual complex containing the sensations T, C will follow or be provoked by B, or we infer the second from the first. If the volition B was not allowed to occur, we say that C would have followed it.[5]

If now we have a new complex of sensations A_x that also contains T, we infer, judge, conjecture or opine[6] that C will follow (or be generated by B). If this actually occurs, our surmise is confirmed by experience, if not we are surprised, a new memory joins our old ones, our inner picture of actuality is completed, corrected, adapted. We form volitions that call to mind memories and produce sensations accelerating that process. We look for what A_x contains, for what differentiates it, for the cause: we investigate and experiment.[7]

All these processes can be further complicated in countless ways. To form a picture of what in a given case we should expect, further complicated activities of the will (constructions, calculations) may be required. The picture can be so comprehensive that we may use it under the most varied conditions to construct a successful solution. If we experiment with the pictures themselves, calling to mind by volitions their common features and their differences, while seeking to construct a successful solution in cases that differ from observed ones, then we may be said to speculate. The result will have to be tested by experience, as with simple conjectures.

Of opinions sufficiently often confirmed by experience we say that they are certain and that what they express belongs to our knowledge. To construct thought-pictures we constantly need designations for what is common to various groups of phenomena, thought-pictures or intellectual operations: we call them concepts.

If, in the above example, C follows upon a complex A_y, as yet unfamiliar to us, we say that we have explained the latter as soon as we find T in it; or, if all A are so far still unknown to us, when we have observed them and found T in all of them, including A_y (explanation of Arago's experiment by Faraday's discovery of induction currents).

How then do we come to distinguish certain sensations as our own and others as other people's? The series of sensations we call our own is much more directly linked with the formation of our memory pictures than are the sensations of others. Every one of our own sensations arouses a memory picture even if only fleetingly, whereas an alien sensation acts on our own memory pictures only if it affects sensations of our own. Our world picture would be ideally perfect if for each of our sensations we had a sign and furthermore a rule by which to construct from these signs the occurrence of all our future sensations and the way they depend on our volitions. If for this it is enough to predict our own sensations, which is indeed the only thing we can test, while the sensations of others can affect our world picture only via our own, how do we ever come to form signs for the sensations of others?

As to that, even our childhood observations are informative. With certain processes attended by certain complexes of sensations (the approaching of the visual picture of my hand to that of a flame) we experience new and sometimes violent sensations that result in volitions in turn affecting the perceptual complexes (we see the picture of the hand withdrawing). The very similar visual picture of another's hand behaves in a perfectly analogous manner.

In speech, certain volitions produce certain movements of the lips (visible for example in a mirror) and aural sensations. On other visual images very similar to the mirror image of our own head we see the same labial movements and at the same time experience the same aural sensations.

We said that the purpose of thinking was the foundation of rules for our ideas such that future sensations are thereby announced in advance. This aim is attained in large measure if we apply the experience gained from perceptual complexes concerning our own bodies also to the interaction of those very similar complexes that relate to the bodies of others. The laws according to which our own sensations run their course are familiar to us and lie ready in memory. By attaching these same memory pictures also to the perceptual complexes that define the bodies of others, we obtain the simplest description of these complexes.

Another's hand behaves just as though on touching a fire a feeling of pain occurred, another's lips as though volitions were acting on them. Of these alien sensations and volitions we have not the least knowledge, but know only our own ideas of them, with which we operate as with those of our own sensations and volitions, thus obtaining useful rules for constructing and predicting the course of our sensations relating to the bodies of others. Thus our conception of the sensations and volitions of others is merely the expression for certain equations always holding between the behaviour of our sensations relative to our own and other people's bodies; it is in a pre-eminent sense what we call an analogy (albeit not a mechanical but a psychological one).

What then is the sense of asserting that these alien sensations and volitions exist as much as my own? Does this not add some hypothetical and unprovable element to the facts? Does it not contravene the task of my ideas as merely to describe the facts?

If by considerations such as these one imagines to have proved that matter is merely the expression of certain equations between complexes of sensations, so that the assertion that matter exists in the same way as our sensations exceeds our task of merely describing, it would be well to remember that this would be proving too much; for in that event the sensations and volitions of all others could not be on the same level as the sensations of the observer, but would have to be taken as merely expressing equations between his own sensations.

Let us analyse this further: in accordance with our introductory remarks, we have not proved anything but merely described; nor, in the sequel, shall we be able to prove anything, but merely to develop certain views psychologically.

The question whether the unicorn or the planet Vulcan exists in the sense in which the stag or the planet Mars exists has naturally a quite definite sense, which is clear from our empirically known relation to the second two items. If, however, someone were to assert that only his sensations existed, whereas those of all others were merely the expression in his mind of certain equations between certain of his own sensations (let us call him an ideologist), we should first have to ask what sense he gives to this and whether he expresses that sense in an appropriate way. Evidently he would still have to denote alien sensations with the same signs analogously arrayed with which he denotes his own; subjectively it would be indifferent to him whether he said that those sensations belonged to others who exist or to others whom he imagines, since for him others are indeed only something imagined. But since we use the verb 'not to exist' when we find that expectations expressed by certain mental signs are not confirmed by experience (I thought, erroneously, that my friend had a brother, now I learn that no such person exists), it would be inappropriate to say that all others, save the person here thinking, did not exist.

The ideologist's assertion ought much rather to run as follows: I use the term 'sensation' or 'act of will' as a thought symbol in three ways: firstly, to represent sensations and volitions immediately given to me; secondly, if I find it useful to link the same terms according to the same laws in order to describe certain regularities between my perceptual complexes (I distinguish this second mode of employment by saying that the terms are signs for sensations and volitions of others), and thirdly, either if previously I wrongly thought the terms would be useful for representing such regularities, or, without ever believing this and for quite different reasons (practising, playing), I combine terms that are quite analogous to those for my sensations and volitions according to laws that are quite analogous: these I call terms for sensations and volitions of non-existing people that I merely imagine.[8]

In this form, however, the ideologist's assertion no longer differs from the ordinary way of expressing these matters. The second point expresses the enormous subjective difference that for me exists between myself and others, but we have so far totally refrained from any judgement as to objective existence.

As with ideology, so with the (idealist) assertion that matter is merely the expression for equations between perceptual complexes.[9]

Since we have reserved the term 'not to exist' for the satellite of Venus, the philosopher's stone and so on, it would evidently be inappropriate to say that matter does not exist. Thus all that remains is the assertion that what we call processes in inanimate nature are *for us* mere ideas for representing regularities of certain complexes of our sensations. In this respect processes in inanimate nature are thus on the same level as the sensations and volitions of others, whereas subjectively our own sensations are much closer to us; but the ideas of inanimate objects that subsequently turn out as incorrect or that have been formed from the start with the proviso that we have no such complexes of sensations as they represent, all these are on the same level as the idea of non-existing people.

I hope that what I have developed so far is perfectly clear. We do not perceive the sensations of others. However it does not complicate but simplify our world picture if in thought we attach them to the complexes of sensations that we call other people's bodies. Therefore we denote these alien sensations with analogous mental signs and words to those for our own (we imagine them), because this gives us a good picture of the course of many complexes of sensations and thus simplifies our world picture.

To express the fact that these are imagined sensations, we say they are not ours but those of others. If we ourselves do not experience the complexes of sensations whose representation would be facilitated by these alien sensations, then we call these latter non-existent. A child may well believe that dolls, trees and so on have sensations too; we do not ascribe sensations to these objects because this would complicate and not simplify our world picture.

Analogously to the sensations of others, processes in inanimate nature likewise exist for us merely in imagination, that is we mark them by certain thoughts and verbal signs, because this facilitates our construction of a world picture capable of foretelling our future sensations in inanimate nature. Processes in inanimate nature in this respect are thus just like the sensations of others, and inanimate objects themselves like those others, except that the signs and laws of their conjunction are rather more different from those used in representing our own sensations. 'An inanimate object either does or does not exist' thus has the same significance as 'a person either does or does not exist'. It would therefore be a total mistake to believe that in this way one had established that matter is more of a mental entity than another person is.

It is true enough that we can build up our world picture only from our sensations and volitions but of all our sensations only the one or few that we momentarily have are directly given to us. It would therefore be an error to think that the memory of having had a sensation is certain proof that it has existed. Children of three often do not yet distinguish memory from phantasy. Those troubled with nocturnal emissions, remembering an incident in the morning, may be uncertain whether it was real or dreamt. If our mental life were never more regular than in dreams, we should at best attain certain laws concerning the change of ideas but never the concept of something existing beside ourselves.

Since moreover very faint memories shade imperceptibly into oblivion, and here and there mere chance calls things to mind that we might under other circumstances never have remembered, we are certain to have had countless sensations, ideas and volitions which we are absolutely unable to recall. It would however be clearly impracticable to fix on a certain degree of imprecision in the memory of a process and from that point abruptly to say that the process never occurred; therefore we must simply denote as existing much that has no direct link with our present-day thought. Moreover we see that many sensations occur in spite of all volitions by which we strive to prevent them, so that there is also something independent of our will. Thus there certainly exist processes that are independent of our present thought and volition, whose existence is 'objectively correct' but not cognizable by us. What is present in our memory is different at different times. In this way we first obtain the concept of objective existence as something independent of momentary memory.

Besides there is another element at work. One of the most important ways in which our world picture develops further is through what others convey to us and what we tell them. In this everyone will naturally distinguish himself as speaker (subject) from those spoken to (objects) and at first adopt the (subjective) point of view we have assumed till now.

It will be appropriate to call our concept of existence and non-existence as discussed so far the concept of subjective existence or non-existence.

Now it would doubtless be inappropriate to address people thus: "your sensations are by no means equivalent to mine. Whereas I am directly conscious of mine, that which I call yours are for me a thought symbol for certain regularities of my own sensations: It is only because certain of my complexes of sensations that I call your bodies consistently change as though they were driven by volitions quite analogous to those I exert on other such complexes of mine (my own body), that I must proceed towards you as your apparent volitions proceed towards me". One would be constantly repeating words that are of no concern to others, that is of no or only undesired impact on those complexes of mine that I call their bodies.

Language must therefore use some other terminology that is equally appropriate for all persons; "we must adopt the objective point of view", as the phrase goes. It turns out that the concepts we linked with 'existing' and 'not existing' largely remain applicable unchanged. Those people or inanimate things that I merely imagine or conceive without being forced to do so by regularities in complexes of sensations do not exist for others either, they are 'objectively' non-existent.

On the other hand, sensations that I assume, without perceiving them, as alien (that is serving to explain regularities belonging to my own), these sensations divide into ones belonging to many other people of whom each is related to his as I to mine.

If therefore I am to make myself understood, I must adopt a language in which all exist on the same footing ('objectively'). This adherence to the language of others which is given to me in experience (because learnt) I call the objective point of view, in contrast to the subjective one so far described.

Since my waking sensations are the only building blocks of my thinking I must start from them; thus, sensations that all my memories agree were waking ones I must denote as that which primarily exists, if all thought is not to stop. For the sake of linguistic homogeneity I must denote the sensation of others in the same way. The criterion that all men should judge alike as regards existence and non-existence applies equally to the phenomena of inanimate nature. However, here we do not have the argument that some extraordinarily similar phenomena are directly given to me so that I must think of them primarily as existing; everybody could therefore agree to distinguish processes in inanimate nature from psychological ones by the fact that they denote the former as not objectively existing. This would indeed be inappropriate if only for the fact that for myself subjectively others and inanimate things that exist are on the same level while non-existing people and things among themselves play the same role, so that for subjective existence the psychological and inanimate are equal;[10] nevertheless, this evidently was the reason why many philosophers held the view that the animate and sensing alone existed, while the inanimate existed only when being perceived by an animate perceiver, whereas in fact another animate being too exists for me only when I perceive it: not only matter but also other people are for me (if I do not accommodate myself to their alien language) mere mental symbols, just an expression of equations between complexes of my sensations.

It would of course be absurd to prove or disprove the objective existence of matter. Rather, it will be a case of giving further reasons why it would be inappropriate constantly to remind ourselves of the fact stated earlier, namely that we denote matter as not objectively existing, although we should always remain clearly aware of this fact.

If somebody regards it as obvious a priori that matter does or does not exist, this, in the absence of some prejudice, can be considered only as expressing the subjective conviction that either one or the other designation would lead to quite ludicrous complications. Such a subjective conviction can of course rest on error too, as when a child cannot imagine a world picture other than one in which everything has the same sensations as himself.

To fix the concept of objective existence we earlier appealed to the common judgement of all. One might of course imagine other man-like beings on other planets or beings of higher intelligence, whose coincident judgement would definitively determine objective existence. However, little would be gained by this; we must therefore return to our own experience.

The reason why we denoted the sensations of people other than the thinking subject as objectively existing, was only their perfect analogy with that subject's sensations, which have the first claim to be so described. We shall therefore still have to examine whether processes in inanimate nature have as many analogies with psychological ones, or whether so sharp a line can be drawn between the two that the former can be described as objectively not existing.

To start with, the sensations of higher animals are so perfectly analogous to human ones that we must of necessity ascribe objective existence also to them. Where then is the boundary? One does indeed hear occasional doubts whether insects or divisible animals like certain worms have sensations, but a sharp boundary where sensing stops cannot be given. In the end we reach organisms that are so simple that their world pictures and thoughts are zero. If we are not suddenly to deny existence to the sensation of animals below a certain level, which would be quite inappropriate, then we must ascribe existence also to this unthinking organised matter, in which sensation can hardly be discovered, but this in turn runs through continuous gradations as far as the level of plants. But then it would seem to me an unjustified and inappropriate jump to deny existence to unorganised matter.

If this were the only argument for objective existence of the inanimate, then a thorough adherent to this point of view might conceive the notion of suggesting the assumption of different degrees of existence which finally sinks to zero for what is inanimate. However such a mode of expression would again be decidedly inappropriate. In the first place we have in any case already got concepts to denote the same fact: we say that the clarity of awareness gradually sinks to zero. Secondly we have already fixed the concept 'existence' in such a (subjective) sense that it does not admit degrees of comparison (another person that exists and one that does not, two moons of Mars exist, a moon of Venus does not exist); the denotation must always be so chosen that we can operate with the same concepts in the same way under all circumstances, just as the mathematician defines negative or fractional exponents in such a way that he can operate with them as with integral ones.

Words, and therefore concepts, we can form as we wish. Somebody once took the trouble to demonstrate to me that a high-school teacher is actually a professor and that therefore our law alone, which gives him this title, is just. I have the same feeling when a word like 'exist' is singled out from language and, without fixing its sense, people start racking their brains as to what exists and what does not.

Progress in thinking must much rather be sought by eliminating all such mistaken forms of inference and concepts, which, experience tells us, do not advance but mislead and even entangle us in contradictions. These forms of inference and concepts always arise when originally appropriate modes of thought are transferred to cases where they do not fit. Thought must be continuously further adapted and the sense of words be ever more appropriately fixed, which in the case of the simplest concepts cannot occur through definition but only by reference to familiar experience.

We see moreover that those series of sensations and volitions that we call single individuals always soon come to break off again, that individual people die, whereas the matter to which those mental phenomena were tied remains. A subjective world picture that construes matter as merely expressing equations between the complexes of human sensations thus starts by trying to imitate the transient and complicated features by means of marks and only later using these pictures to represent simple and more permanent features (of matter). It construes the pyramids of Egypt, the Acropolis of Athens as mere equations existing between the sensations of generations through thousands of years.

Alongside this it must surely be possible to have a simpler (objective) world picture that starts from the simple and represents the transient by means of laws that govern the more permanent features. Pursuing our mental picture consistently, that is according to the rules that have always led to confirmation by experience, we reach the conclusion that the planet Mars is of similar size to the Earth, that it has continents, oceans, snowcaps and so on, indeed it seems not at all impossible to us that on planets of other suns there are the most splendid landscapes without their ever producing sense impressions on any animate being.

For us subjectively the expression of this is of course only a minor internal activity of imagination or a few spoken sentences that have nothing in common with the immense cosmic processes in question. These mental or verbal signs have for us no other significance than the possibility of certain geometrical constructions on a reduced scale, a linking of them with series of numbers and some analogies or other with terrestrial landscapes, which in analogous cases on Earth are always confirmed by experience and without which our world picture would be inconsistent and incomplete. From this we infer the possibility of beings similar to ourselves to whom these landscapes mean the same as terrestrial ones to us, with as much justification as we infer that we must have had many sensations that we no longer remember.[11] Here our sensations quite automatically lead us beyond their own fields to detailed and definite ideas of things that are remote from the sphere of our sensations.

Would not the man who considered the Martian landscape from the point of view of equations between the sparse human sensations related to Mars have just as onesided and inappropriate a world picture as he who regards himself alone as existing and others not? For possible Martians would not exist for us either until we could have perceptions relating to them.

We see further that our intellectual activity affects that of another person only when by means of volitions we produce changes in those complexes of sensations that correspond to matter and when these become so related to another's body that we too should receive sense impressions. Nowhere do we find direct equations between our own and another's sensations, all are mediated by matter. It is therefore between changes of matter that we must expect to find the simplest equations.

The intimate connection of the mental with the physical is in the end given to us by experience. By means of this connection it is very likely that to every mental process there corresponds a physical process in the brain, that is, there is an unambiguous correlation; and that the brain processes are all genuinely material, that is, are representable by the same pictures and laws as processes in inanimate nature. In that event, however, it would have to be possible to predict all mental processes from the pictures that serve to represent brain processes. Thus all mental processes must be predictable from the pictures used for representing inanimate nature without change of the laws that govern it. Let us give the name A to the view that this is correct.

All these circumstances make it extremely likely that an (objective) world picture is possible in which the processes in inanimate nature play not only the same but even a much more comprehensive role than mental processes, which latter are then related to the former only as special cases to general ones. Our aim will not be to establish the truth or falsehood of one or the other world picture, but we shall ask whether either is appropriate for this or that purpose while we allow both pictures to continue alongside each other.

If so far we have started with the genesis of our world picture, constituting it purely synthetically, let us now adopt the opposite course in order to represent the objective world picture, a path which is as a rule the most appropriate where what counts is the laying bare of concepts as precisely as possible. We merely give the most easily grasped rules for constructing this world picture without bothering how we subjectively came by these rules: the only justification for the world picture is then seen in its agreement with the facts. What previously came first will now come precisely last.

The brain we view as the apparatus or organ for producing word pictures, an organ which because of the pictures' great utility for the preservation of the species has, conformably with Darwin's theory, developed in man to a degree of particular perfection, just as the neck in the giraffe and the bill in the stork have developed to an unusual length. By means of the pictures by which we have represented matter (no matter whether the most suitable pictures will turn out to be those of current atomism or some others), we now try to represent material brain processes and so to obtain at the same time a better view of the mental and a representation of the mechanism[12] that has here developed in the human head, making it possible to represent such complicated and apposite pictures.

The moment we subscribe to the view A, we must suppose that the pictures and the laws that serve the representation of processes in inanimate nature will suffice unambiguously to represent mental processes too; we say, in brief, that mental processes are identical with certain material processes in the brain (realism). It has often been held that this is impossible. Whether we are entitled to hold this view we can naturally test once again only from what is given in experience.

From experience we know that every sensation somehow differs from every other and that some resemble each other more and others less, so that the former have more in common and the latter less; besides we know in what order they occur in time. As to quality, the finer or cruder, material or non-material nature of sensations we know nothing from direct experience. Hence I cannot grasp why people say that we sense (or know a priori or are immediately aware or whetever else) that sensations are simple or qualitatively different from processes in inanimate nature or even that they are finer, loftier and so on. Indeed, some have thought they sensed that the whole human ego was something simple. On the contrary, it is precisely the protean variability in the nature of different sensations and their almost indefinable similarities that makes it likely that their course cannot be represented by the most simple mental pictures but only by very complicated ones, like the various physical and chemical processes in the brain.[13] Again we want to express just this and no more when we say that thoughts are certain processes in the brain or perhaps the interaction of certain atoms.

If one says that matter or even atoms sense, one has obviously expressed oneself quite incorrectly. Rather one must say that it is not inconceivable that the laws of change in sensation are most accurately representable by means of the picture of material (physical, chemical, electrical) processes in the brain.

The most complicated systems of material bodies whose mode of working is more or less transparent to us, are perhaps objects like a watch or a dynamo machine. We therefore believe that if our psychological processes were completely representable by means of pictures of material processes in the brain, they would have to be just as dead and unengaged as these machines. That evidently is the reason why to many this view seems barren and comfortless. Unjustifiably so, in my opinion, for the genesis of strong feelings of pleasure and pain is precisely what Darwin's theory can explain, since such feelings are necessary to generate reactions of the intensity required for the preservation of the species. The great intensity, variety and wealth of intellectual and emotional life can surely not be caused by the processes in question being qualitatively finer and loftier than those in dead machines but only by their being richer and more varied and by our own egos' belonging to the same kind of beings. Since no one will doubt that intellectual functions too work according to quite definite laws, I could surely not find it discouraging that these laws were identical with those that govern equally complicated material processes. It is simply that, for our subjective feelings, that is fine and lofty which advances and raises our species: objectively these concepts do not exist. If therefore material processes can be just as varied and complicated as our mental ones, and there is no reason for doubting it, then I do not see why the assertion that our mental activities are completely representable by the thought picture of material processes in the brain should impair their fine and lofty character or somehow interfere with our passionate interest in them. We know that a watch does not sense, that is, by means of such a simple mechanism we cannot represent anything remotely similar to sensations. But what is it supposed to mean when people say that from the qualitative differences between our sensations and material processes it follows that the course of the former could never be represented by any combination, however complicated, of mental pictures that at the same time represented for us the processes of inanimate nature. If one says that the inanimate world is material, extended and so on, one merely means that it is representable by the conceptual pictures of geometry and mathematical physics. If therefore one asserts on the contrary that sensations are non-material, unextended and so on, one has merely assumed beforehand what was to be proved, namely that they cannot be represented by combinations of these pictures however complicated. The fact that we have not to date succeeded in representing the genesis of sensations by means of complicated pictures taken from physics and chemistry surely does not prove this to be impossible in principle? Our judgement about the representability of a group of sensations by means of certain pictures naturally remains completely unstable and indefinite so long as the representation is not completely successful to the minutest detail. The pictures of geometry and mechanics were set up in order to represent ordinary phenomena of equilibrium and motion and this has succeeded so well that we do not doubt the possibility of representing all phenomena of the field in question. All other purely physical processes cleave so intimately to material bearers that our need to enlist the pictures of geometry and mechanics for their partial explanation is beyond doubt. Whether these pictures are sufficient everywhere is however a question about which views still differ greatly. Even the phenomena of heat sometimes have features that seem at least at first blush to be not just spatio-temporal but of another kind, let us say qualitative changes of bodies; and whereas some physicists believe that they can best be represented by the picture of motions of the smallest parts, others think it unlikely. Even more doubt exists as regards electro-magnetic phenomena, radiant energy and chemistry. Indeed, one even hears the view that these last phenomena would require for their representation an extension even of the pictures of geometry. Thus even purely physical facts are by no means all of the same kind. Yet who would assert that this constitutes strict proof of qualitative gaps so great as to make representation by mechanical pictures certainly impossible in principle?

Mental phenomena may well be much more remote from material ones than thermal or electric from purely mechanical ones, but to say that the two former are qualitatively while the latter three are only quantitatively different seems to me mere prejudice.

If we make our previous assumption A, that to every mental process unambiguously corresponds a certain brain process and these last are genuinely material, that is representable by pictures and laws that serve for the representation of processes in inanimate nature, then on the contrary the genesis and course of mental phenomena would have to be unambiguously definable (that is representable) by these laws.

Imagine there could be a machine[14] that looked like a human body and also behaved and moved like one. Inside it let there be a component that receives impressions of lights, sound and so on, by means of organs that are built exactly like our sense organs and the nerves linked with them. This component is further to have the ability of storing pictures of these impressions and by means of these pictures so to stimulate nerve fibres that they produce movements that are totally similar to those of the human body. Unconscious reflex movements would then naturally be those whose innervation did not penetrate so deeply into the central organ as to generate memory pictures there. It is said to be a priori clear that this machine behaves externally like a man but does not sense. It would indeed retract the burnt hand just as quickly as we do, but without feeling pain. I think that people say this merely because one visualizes only a clock and not our present complicated machine, just as people uninformed about physics often tell me that they find it (we should say a priori) clear that even in space one must still know what was up and down, or that one must feel it if the Earth were turning. These people have simply failed to imagine themselves in outer space and to conceive of cosmic conditions.

However compelling such judgments for the biassed, they prove nothing. In our fictitious machine every sensation would exist as something separate. Similar sensations would have much in common and dissimilar ones less. Their course in time would be that given by experience. Of course no sensation would be simple, each would be identical with a complicated material process, but for one who does not know how the machine is built, sensations would again not be measurable by length and measures, he could no more represent them by spatial and mechanical pictures than we can our own sensations. However, nothing more is given by experience. Thus everything we are empirically given of the mental would be realised by our machine. The rest we arbitrarily add in thought or so it seems to me. Like any other person, our machine would say that it was aware of every existence (that is, it had thought-pictures for the fact of its existence). Nobody could prove that it was less aware of itself than a human. Indeed, one could not define consciousness in some manner such that it applied less to the machine than to men.

In the last few sentences we have returned completely to the one-sided position, and to the channels of the old terminology, which can naturally always be applied as long as one harbours appropriate ideas in using it. To exclude misunderstandings we repeat that these last considerations are merely to show how one can build oneself a world picture starting from a certain point of view. This in no way involves the question of the ideal nature of the human spirit. Indeed, everything remains as it was. We merely declare it possible that the same thought symbols and laws through which we obtain the best pictures of processes in inanimate nature might in more complicated combinations equally afford us the simplest and clearest pictures of mental processes.

If one thus adheres to this our view A, processes in inanimate nature differ so little in quality from animate ones, that it is impossible to draw any boundary and it would be impracticable to ascribe objective existence only to sensations but not to processes in inanimate nature. It would much rather be questionable whether dreamt sensations, or merely recollections of them on waking, existed objectively, a question which might however be decided in terms of the physiology of the brain.

The synthetic description of the genesis of thoughts naturally remains the following: to begin with we construct thought pictures of the sensations of which we are immediately aware; then we come to thought symbols for those regularities of our perceptual complexes that lead to the idea of matter. In thus representing material processes in the brain (which we might indeed one day be able to observe objectively, for example by means of X-rays), we hope to attain a better quantitative survey of the mental processes that were our starting point. But would this not amount to proving that what we see by means of X-rays was something quite different from our sensations? Not at all: we should now have demonstrated a new connection between various sensations, namely those that we have long since known and certain visual pictures that only arise when we look at a screen that intercepts X-rays that have passed through our heads.

If, however, one wishes to reject view *A*, then one must assume either that not all processes in the brain are representable by the pictures and laws that serve for representing inanimate nature, or that there are mental processes not representable by these pictures and laws to which correspond no brain processes, a position made unlikely by experience but not absolutely refuted. But then the gap between animate and inanimate would indeed grow deeper. However, idealism would still be confronted with the difficulties already mentioned, for example the bridging of this gap by a gradual transition from animate to inanimate, the dominant role the inanimate would have to play in any world-picture, in contrast with which the mental will appear merely as a kind of appendix. However the ideas of the thinker himself cannot be thus left out if the world picture is not to vanish completely. Persons close to him also have great influence on his world picture and all preceding generations have supplied the preconditions of his own development. Yet all living beings on all celestial bodies other than the Earth, nine-tenths or more of everything animate that ever existed on Earth could be thought of as never having existed almost without disturbing the world picture. Or one might think everything living on most parts of the Earth being suddenly annihilated without our noticing it at first, whereas a sudden annihilation of a part of the Earth or Sun or even the Moon would throw everything out of gear.

The idealist compares the assertion that matter exists as much as our sensations with the opinion of the child that a stone feels pain when struck. The realist compares the assertion that one could never imagine how the mental could be represented by the material let alone by the interaction of atoms with the opinion of an uneducated person who says the Sun could not be 93 million miles from the Earth, since he cannot imagine it. Just as ideology is a world picture only for some but not for humanity as a whole, so I think that if we include animals and even the universe the realist mode of expression is more appropriate than the idealist one.

Thus from insights or experiences already gained one can indeed demonstrate new aspects of them, but the simplest preconditions of all experience and the laws of all thought one can, I think, at best describe. Once admit this and all contradictions vanish that one previously met in the attempt to answer certain questions, for example whether complexes of unextended atoms make up extensions or whether they can even sense, whether we can come to know the sensations of others or the existence of inanimate beings, whether matter and spirit can interact, whether both run parallel courses without interacting or even whether only one or the other exists. One sees that one did not know what one was really asking.

Here, too, belongs the question of the existence of God. It is certainly true that only a madman will deny God's existence, but it is equally the case that all our ideas of God are mere inadequate anthropomorphisms, so that what we thus imagine as God does not exist in the way we imagine it. If therefore one person says that he is convinced that God exists and another that he does not believe in God, in so saying both may well think the same thoughts without even suspecting it. We must not ask whether God exists unless we can imagine something definite in saying so; rather we must ask by what ideas we can come closer to the highest concept which encompasses everything.

NOTES

1 Contradictions (for example we cannot conceive of bodies being really infinitely divisible, nor yet of an extended body as arising from a finite number of points) can lie only in ways of denoting and are thus a sign that these have been inappropriately chosen. Experience cannot contradict itself, for even if its laws were to change completely, ways of denoting would have to adapt to the new laws.

2 This is the law of causality, which we are thus free to denote either as the precondition of all experience or as itself an experience we have in conjunction with every other.

We can infer from experience that in lotto every move is equally likely. For this reason we have constructed the calculus of probability in such a way that, according to its laws, even if by chance on some occasion a given number came up more often, this does not make its turning up again on the next move any more likely. People now argue thus: it is a priori equally likely that the sun will or will not rise tomorrow, therefore its rising hitherto does not make its rising tomorrow any more likely. To this we must object that an a priori equality of the likelihoods of either event is just as senseless as an a priori knowledge of either, so that according to experience the probability laws of lotto are here not applicable.

[3] The purpose of the whole arrangement is to bring about what is useful to the individual or species and to ward off what is harmful.

[4] It would be quite wrong to infer from this that one must not punish actions that harm the community. One must punish them, that is produce in the criminal and in others memories that will in future prevent the undesired action from occurring. However, only voluntary actions should be punished, since involuntary ones are not affected by memory images.

[5] In the same way we can infer the past. If B had happened, then in the past C would have followed; or another example: I remember once having had a complex of sensations A of which I know that C will always follow it, from which I infer that C did follow on that occasion too, even if I no longer directly remember that it did.

[6] Surmise and opinion are uncertain, inference is almost certain, while judgement refers to the appropriateness of our own ways of denoting or of actions, which I do not in the least intend to discuss further here.

[7] Cf. Mach, *Prinzipien der Wärmelehre*, Leipzig 1896, pp. 386, 416, and elsewhere.

[8] I assume the existence of a man in earlier times (in history) in order to explain, that is represent in thought, accounts about or remains and extant traces of his former activities.

[9] If from this assertion (idealism) one infers that no property of matter (for example that it must consist of immutable particles or that all phenomena must be representable in terms of phenomena of motion) can be recognized as a priori, then I will naturally assent to this demand at once. However, this inference does not exclude our calling matter something existing. For example sensations too are precisely something changeable although they are what is given as existing before anything else.

[10] That is why the rules for handling the concept of objective existence become most conformable to the corresponding rules for handling the concept of what we have called subjective existence, if we denote matter as objectively existing, and that is a main reason why calling it so is appropriate.

[11] It is conceivable that a mental picture, for example atomism, becomes so complicated in its further development that the time available to all mankind will never suffice for developing the picture further still. In that case the assertion that the picture if further developed could represent a large part of the world still has sense even if no practical significance.

[12] The term mechanism is of course not meant to prejudge whether the laws of current mechanics must suffice to represent it.

[13] That is, if the concept of the continuum is properly understood, an interplay of its atoms, by which of course we must not imagine material points but perhaps vectors or whatever. Nor do the atoms necessarily have to be inmutable (cf. *Wien. Ber.* **105**, Nov. 1896; and the essay on p. 41 above).

[14] By a machine I naturally mean merely a system built up from the same constituents according to the same laws of nature as inanimate nature, but not one that must be representable by the laws of current analytical mechanics; for we are by no means sure that the whole of inanimate nature can be represented by these latter.

ON THE DEVELOPMENT OF
THE METHODS OF THEORETICAL PHYSICS
IN RECENT TIMES*

In earlier centuries, science advanced steadily but slowly through the work of the most select minds, just as an old town constantly grows through new buildings put up by industrious and enterprising citizens. In contrast, our present century of steam and telegraphy has set its seal of nervous and precipitate activity on scientific progress too. Especially the development of natural science in recent times resembles rather that of a modern American town which in a few decades grows from a village into a city of millions.

Leibniz has rightly been called the last man who still could unite the entire knowledge of his time in one single head. More recently there has indeed been no lack either of men who amazed us by the enormous scope of their learning. Let me mention only Helmholtz who had attained equal mastery in four fields: philosophy, mathematics, physics and physiology However, these were still only a few more or less related branches of human knowledge as a whole, which reaches very much further.

The consequence of this vast and rapidly growing extent of our positive knowledge was a division of labour in science right down to the minutest detail, almost reminiscent of a modern factory where one person does nothing but measure carbon filaments, while another cuts them, a third welds them in and so on. Such a division of labour certainly helps greatly to promote rapid progress in science and is indeed indispensable for it; but just as certainly it harbours great dangers. For we lose the overview of the whole, required for any mental activity aiming at discovering something essentially new or even just essentially new combinations of old ideas. In order to meet this drawback as far as possible it may be useful if from time to time a single individual who is occupied with the work of scientific detail should try to give a larger and scientifically educated public a survey of the development of the branch of knowledge in which he is working.

This involves no small difficulties. The almost endless chain of infer-

* *Populäre Schriften*, Essay 14. Address to the meeting of natural scientists at Munich 22 September 1899.

78 FROM 'POPULÄRE SCHRIFTEN'

ences or single experiments of which any result is the goal can be seen and easily grasped only by one who has made it his life's task to roam through precisely these chains of ideas. Moreover, in order to make expression shorter and clarity easier it has been found highly useful everywhere to introduce a large number of new terms and scientific words. The speaker for his part must not exhaust his listeners' patience by explaining all these new concepts before reaching the subject proper of his talk, yet without them he will find it hard and awkward to make himself intelligible. Besides, making an account popular must never be regarded as the main object. This would lead to making inferences less strict and abandoning that exactitude which has supplied natural science with an epirhet that is a matter of no small pride. If therefore I have chosen as my present topic a popular account of the development of theoretical physics in recent times, I was well aware that my goal cannot be attained in the degree of perfection with which it is before my mind, but that I shall be able to do no more than give a rough outline of the most generally important aspects, while occasionally having to offend by mentioning well-known facts for the sake of the requisite completeness.

The main cause of rapid scientific progress in recent times lies undoubtedly in the discovery and perfecting of an especially suitable method of research. In the experimental field this method often simply continues working automatically, and the enquirer needs only to go on supplying fresh material as it were, just as a weaver puts fresh yarn on his mechanized loom. Thus a physicist needs only to continue to test new substances for viscosity, electric resistance and so on, repeating these measurements at the temperatures of liquid hydrogen and of Moissan's furnace, and similarly with many tasks in chemistry. Of course it still requires a fair measure of ingenuity to discover in each case what are the experimental conditions under which these things can be done.

It is not quite so simple with the methods of theoretical physics, but there too we can in a sense speak of an automatic running-on.

This eminent importance of the right method explains why men soon started to think not just about things but also about the method of our thinking itself; thus arose the so-called theory of knowledge, which, in spite of a certain tang of old-style metaphysics now discredited, is highly important to science.

The further development of scientific method is so to speak the skeleton that carries the progress of science as a whole. For that reason I shall in what follows make the development of methods the centre of discussion, weaving in the results obtained merely to illustrate the methods. Results are by their very nature easier to grasp and better known whereas it is precisely the way method interconnects them that needs illustration most.

It is especially attractive to follow up a historical account with a look at scientific developments in a future that, in view of the short span of human life, none of us can experience. As to that let me confess at once that all I have to say will be negative. I will not be so rash as to lift the veil that conceals the future, but I will offer reasons that should be apt to warn us against certain hasty conclusions as regards future scientific developments.

A closer look at the course followed by developing theory reveals for a start that it is by no means as continuous as one might expect, but full of breaks and at least apparently not along the shortest logical path. Certain methods often afforded the most handsome results only the other day, and many might well have thought that the development of science to infinity would consist in no more than their constant application. Instead, on the contrary, they suddenly reveal themselves as exhausted and the attempt is made to find other quite disparate methods. In that event there may develop a struggle between the followers of the old methods and those of the newer ones. The former's point of view will be termed by their opponents out-dated and outworn, while its holders in turn belittle the innovators as corrupters of true classical science.

This process incidentally is by no means confined to theoretical physics but seems to recur in the developmental history of all branches of man's intellectual activity. Thus many may have thought at the time of Lessing, Schiller and Goethe, that by constant further development of the ideal modes of poetry practised by these masters dramatic literature would be provided for in perpetuity, whereas today one seeks quite different methods of dramatic poetry and the proper one may well not have been found yet.

Just so, the old school of painting is confronted with impressionism, secessionism, plein-airism, and classical music with music of the future, Is not this last already out-of-date in turn? We therefore will cease to be amazed that theoretical physics is no exception to this general law of development.

Basing themselves on the prior work of many natural philosophers of genius, Galileo and Newton had created a doctrine that must be designated as the beginning proper of theoretical physics. Newton was especially successful in incorporating the theory of celestial dynamics within this structure. He regarded every heavenly body as a mathematical point, as fixed stars indeed appear to be at a first approximation of observation. Between each two of these points there was to be a force of attraction along the line joining them and inversely proportional to the square of their distance. By conceiving a similar force to be acting between any two material particles of any body whatsoever and by applying the laws of motion obtained from observations on terrestrial bodies, he succeeded in deriving from the one law the motions of all celestial bodies, gravity, the tides and all connected phenomena.

Given these great successes of Newton, his successors aimed at explaining all other natural phenomena entirely by his method except for appropriate modifications and extensions. Using an old hypothesis going back to Democritus they conceived of bodies as aggregates of very many material points or atoms. Between any two of these there was, besides Newtonian attraction, another force that was considered to be repulsive at certain distances and attractive at others, whichever seemed most appropriate for explaining the phenomena in question.

Calculation had resulted in the so-called principle of conservation of kinetic energy. Whenever a certain amount of work is done, that is, if the point of application of a force is moved by a certain distance in the direction of that force, a certain amount of motion must arise, whose quantity is measured by a mathematical expression called kinetic energy. Precisely this amount of motion actually appears as soon as the force uniformly acts on all parts of a body, for example in free fall; but progressively less appears if only some parts are affected by the forces and others not, as in friction or impact. In all processes of the latter kind heat is generated instead. Therefore the hypothesis was advanced that heat, which had previously been regarded as a substance, was nothing but an irregular relative motion of the smallest parts of a body with regard to each other, not directly visible since the particles themselves are not, but conveyed to our nervous system and thus producing the sensation of warmth.

It followed from this theory that the heat generated must always be exactly proportional to the kinetic energy lost, a proposition called the equivalence of kinetic energy and heat, and this was confirmed. It was further presupposed that in solid bodies every particle oscillates about a certain rest position and that the configuration of these rest positions determines the solid shape of the body. In liquids molecular motions are so vigorous that the particles creep past each other; vaporisation occurs by total detachment of particles from the surface of bodies, so that in gases and vapours the particles mostly fly off in straight lines like bullets from a gun. This accounts in a natural way for the occurrence of the three states of aggregation of bodies, as well as for many facts of physics and chemistry. From many properties of gases it follows that their molecules cannot be material points. It was therefore supposed that they are complexes of such points, perhaps surrounded by layers of aether.

For, besides the ponderable atoms that make up bodies, the existence was assumed of another stuff made from much subtler atoms, namely the luminous aether whose transverse vibrations afforded an explanation of almost all phenomena of light, which Newton had earlier ascribed to the emanation of special light-particles. Some difficulties did indeed remain, such as the total absence of longitudinal waves in the luminous aether, which in all ponderable bodies not only occur but actually play the main role.

Our factual knowledge of electricity and magnetism was enormously increased by Galvani, Volta, Oerstedt, Ampère and many others, and was brought to a certain finality by Faraday. The latter, using rather limited means, had found such a wealth of new facts that it long seemed as though the future would have to confine itself merely to explaining and practically applying all these discoveries.

The cause of electromagnetic phenomena had long been conceived as special electric and magnetic fluids. Ampère succeeded in explaining magnetism by means of molecular electric currents, which made the assumption of a magnetic fluid superfluous, and Wilhelm Weber perfected the theory of electric fluids by completing it in such a way that all electromagnetic phenomena known up till then could be explained. To this end he conceived of electric fluids consisting of the minutest particles just as ponderable bodies do and the luminous aether, with electric forces between them quite analogous to those between material particles, with the unessential modification that the forces between any two electric particles were further to depend on their relative velocities and accelerations.

Whereas then in the first stages one had assumed beyond tangible matter a caloric substance, a luminous substance, two magnetic and two electric fluids and so on, later one made do with ponderable matter, luminous aether and the two electric fluids. Each of these substances was conceived of as consisting of atoms, and the task of physics seemed confined for ever to ascertaining the law of action of the force acting at a distance between any two atoms and then to integrating the equations that followed from all these interactions under appropriate initial conditions.

This was the stage of development of theoretical physics when I began my studies. How many things have changed since then! Indeed, when I look back on all these developments and revolutions I feel like a monument of ancient scientific memories. I would go further and say that I am the only one left who still grasped the old doctrines with unreserved enthusiasm – at any rate I am the only one who still fights for them as far as he can. I regard as my life's task to help to ensure, by as clear and logically ordered an elaboration as I can give of the results of classical theory, that the great portion of valuable and permanently usable material that in my view is contained in it need not be rediscovered one day, which would not be the first time that such an event had happened in science.

I therefore present myself to you as a reactionary, one who has stayed behind and remains enthusiastic for the old classical doctrines as against the men of today; but I do not believe that I am narrow-minded or blind to the advantages of the new doctrines, which shall receive due justice in the next section of my talk, so far as lies within my power; for I am well aware that like everyone else, I see things subjectively tinged through my own spectacles.

The first attack on the scientific system described was directed against its weakest side, Weber's theory of electro-dynamics. This is so to speak the flower of the intellectual work of that gifted enquirer, who has earned the most immortal merit on behalf of electric theory by his many ideas and experimental results recorded in the system of electrodynamic units

DEVELOPMENT OF METHODS OF THEORETICAL PHYSICS   83

and elsewhere. However, for all its ingenuity and mathematical subtlety, it bears so much the stamp of artificiality, that there can surely never have been more than a few enthusiastic followers who believed unconditionally in its correctness. It was Maxwell who attacked it, while giving fullest recognition to Weber's achievements.

Maxwell's enquiries will concern us here in two respects, the epistemological part and the specifically physical. As regards the first, Maxwell warned against regarding a particular view of nature as the only correct one merely because a series of consequences flowing from it has been confirmed by experience. He gives many examples of how a group of phenomena can be explained in two totally different ways, both modes of explanation representing the facts equally well. Only on adding new and hitherto unknown phenomena does the advantage of one method over the other reveal itself, though the former may have to give way to yet a third after further facts have been discovered.

Whereas it was perhaps less the creators of the old classical physics than its later representatives that pretended by means of it to have recognised the true nature of things, Maxwell wished his theory to be regarded as a mere picture of nature, a mechanical analogy as he puts it, which at the present moment allows one to give the most uniform and comprehensive account of the totality of phenomena. We shall see how influential this Maxwellian position was on the further development of his theory. Through his practical successes he quickly helped these theoretical ideas to victory.

We saw that all electromagnetic phenomena then known were explained by Weber's theory, which viewed electricity as consisting of particles acting on each other directly and without any transmission at all distances. Inspired by Faraday's ideas Maxwell now developed a theory starting from the opposite point of view. According to this every electric or magnetic body acts only on the neighbouring particles of a medium that filled the whole of space, and these in turn on their neighbours in the medium until the action has propagated itself to the next body.

The phenomena known till then were equally well explained by both theories, but Maxwell's went much beyond the old theory. According to his theory, as soon as it was possible to produce sufficiently fast electric motions, they produce in the medium wave motions that exactly obey the laws of motion of light-waves. Maxwell therefore surmised that within the particles of luminous bodies there are constant rapid motions of electricity and that the oscillations thereby provoked in the medium are precisely light. The medium transmitting electromagnetic effects thus becomes identical with the luminous aether already previously required so that we may be allowed to call it so, although it must have properties that are largely different if it is to serve for the transmission of electromagnetic effects.

The reason why in earlier experiments on electricity such oscillations could not be observed may perhaps be illustrated as follows. Let us put an open hand on a pendulum at rest gradually raising the pendulum at right angles to the bar and moving it towards the side where it touches and back again and finally remove the hand completely towards the other side. The pendulum, following the hand, makes half an oscillation but does not continue to oscillate because the velocity imparted to it is too small. Another example: theory assumes that on plucking a string one of its points is displaced from equilibrium and then the whole string suddenly is left to itself. As a student I did not believe this but thought that the person who plucks must give the string an extra push, for when I first bent the string outwards with my finger and then removed it quickly in the direction in which the string was to oscillate, the latter remained silent. I overlooked the fact that in relation to the string's vibration I moved my finger much too slowly and therefore actually impeded it.

Just so in previous experiments the electrical states were always transformed at speeds relatively much too slow in comparison with the enormous speed of propagation of electricity. After laborious preliminary experiments whose leading ideas he describes with complete candour, Hertz now found certain experimental conditions under which electrical states are changed periodically at such a rate that observable waves result. Like any discovery of genius, these conditions are extremely simple. Nevertheless I cannot here enlarge even on these simple experimental details. The waves that Hertz thus produced, undoubtedly through electric discharges, are, as Maxwell had predicted, qualitatively not at all different from lightwaves. But what a vast quantitative difference!

As pitch in the case of sound so colour in the case of light is known to be determined by frequency. In visible light the extreme red end has a frequency of about 400 billion oscillations per second, and extreme violet 800 billion. Very similar aether waves had long since been discovered with frequencies of about 20 times smaller than at the red end and 3 times greater than at the violet end. They are invisible to the eye but the former, so-called infra-red waves are recognisable by their thermal effects and the latter, ultra-violet ones by chemical and phosphorescent effects. The waves that Hertz produced by actual discharges had no more than 1 000 million oscillations a second and his successors reached about one hundred times more.

That such slow oscillations cannot be detected by eye is obvious. Hertz proved their existence by microscopically small sparks which they produce in suitably shaped conductors even at great distances. These conductors might thus be described as eyes for Hertzian oscillations. By these means Hertz confirmed Maxwell's theory to the last small detail; and although people tried to obtain oscillations from a theory of action at a distance, everybody soon shed all doubt as to the superiority of Maxwell's theory, indeed just as a pendulum goes beyond its rest position to the opposite side, extremists went so far as to brand all conceptions of classical physical theory as misguided. But of this later; first let us tarry a while longer amongst these splendid discoveries.

Of the various aether waves, familiar already before Hertz, some were long known to pass more easily through one body, some through another. Thus a dilute alum solution lets through all visible but few infra-red radiations, which for their part easily go through a solution of iodine in carbon disulphide that is quite impervious to visible light. Hertzian waves pass through almost all bodies except metals and electrolytes. When therefore Marconi generated very short Hertzian waves at one place and at another many kilometres removed transformed them into Morse signals by means of a suitable modification of the apparatus we just called an eye for Hertzian waves, he really constructed none other than an ordinary optical telegraph, except that he used waves of a frequency of about a tenth of a billion per second instead of about 500 billion per second. This has the advantage that his waves pass unattenuated through mist and even rock. A mountain of pure metal or a cloud of mercury droplets would be just as impervious to them as ordinary mountains and mist to visible light.

The variety of known radiations was further increased by the rightly celebrated discovery of Röntgen rays. These go through all bodies, including metals, though these last and bodies containing metal such as bones, which contain calcium, greatly attentuate the rays. Phenomena of polarisation, interference and refraction, established for all previously discussed radiations, have not yet been observed for Röntgen rays: if they were really incapable of all polarisation, they would have to be long-itudinal, if they be waves at all; but we must even allow that they might be incapable of interference too and therefore simply not waves at all which is why we speak cautiously of rays and not waves. If we discovered a body that polarised them, this would be evidence for their being quali-tatively similar to light, but they would have to have a very much shorter period of vibration than even the furthest ultra-violet, or perhaps consist merely in rapid successions of shock-waves, although few physicists believe this.

Given this enormous variety of radiations we are almost tempted to argue with the creator for making our eyes sensitive for only so minute a range of them. This, as always, would be unjust, for in all areas only a small range of a great whole of natural phenomena is directly revealed to man, his intelligence being made acute enough to gain knowledge of the rest through his own efforts.

If Röntgen rays were really longitudinal rays of the luminous aether, as its discoverer was from the start inclined to believe, a point that remains unrefuted to this day, we should be confronted by a peculiar case far from unique in science. Classical theoretical physics had a fully worked out view concerning the constitution of the luminous aether. One thing alone, it was held, was still needed to establish its irrevocable correctness, namely longitudinal aether waves; but these could not be found, cost what it might in effort. Now that it has been shown that the luminous aether must have a significantly different constitution since it transmits electric and magnetic effects as well, now that the old view as to its make-up is discarded, we are coming close to the discovery of longitudinal waves in the aether when the time for aspiring to their confirmation is past.

Weber's electro-dynamic theory fared similarly. It is based, as we have seen, on the assumption that the effect of electric masses depends on their relative motion, and just when its inadequacy was definitively proved, Rowland found by a direct experiment in Helmholtz's laboratory that moving charges act differently from stationary ones. In earlier periods one might have felt inclined to regard this as a direct proof that Weber's theory was correct. Today we know that it is not a crucial experiment but that it follows also from Maxwell's theory.

It follows further from a modification of Weber's theory that not only current-carrying conductors but also the currents themselves must be deflected by magnets. This phenomenon, too, after being long sought in vain, was discovered by the American physicist Hall at a time when the adherents of Weber's theory were long past savouring the triumph because of much greater prior defeats.

Such phenomena show how careful one must be if one is to regard a confirmed consequence as a proof for the absolute correctness of a theory. According to Maxwell's view we often find that pictures that have been adapted to nature in many cases automatically remain correct in certain others as well, but this does not prove that they will do so in every case. On the other hand these phenomena show that even an incorrect theory can be useful, if only it contains leads to novel experiments.

The cited discoveries by Hertz, Röntgen, Rowland and Hall established that Faraday had after all left someting for his successors to discover. Many other discoveries followed quite recently, of which we shall here mention only that of Zeeman concerning the influence of magnetism on emitted light and the corresponding effect on the absorption of light. All these phenomena, many of which Faraday was looking for, could not be observed with the instruments of his time. If therefore a genius often achieves the greatest results with the smallest resources, we here see the opposite, that it requires the enormous perfection of present day instruments of observation and experimental technique before the human intellect can achieve certain kinds of result.

Most of the quite novel phenomena here described are so far known only in their basic features. Exploring them as to detail and as to their relations to each other and to the rest of all known phenomena, their embedding into the mechanical physical loom, to use somewhat exaggerated language, all this will open up a future field of work of seemingly boundless extent. The copious practical results achieved from the very start (X-ray photography, wireless telegraphy, radio-therapy) allow us to imagine the practical gain that detailed research will afford, which is ordinarily required before any fruitful results emerge. Theory, however, has been shaken from its complacency in which it thought it had already recognised everything, nor has it been possible yet to gather the new phenomena into as uniform a theoretical structure as the old theory had been. Everything remains rather in a state of indecision and ferment.

This confusion was further increased by various additional circumstances acting together with those already mentioned. For a start we must mention certain philosophical reservations about the foundations of mechanics, stated most distinctly by Kirchhoff. The dualism of force and substance had been unwittingly absorbed into the old mechanics. Force is regarded as a special agent alongside matter, as the cause of all motion; indeed people occasionally quarrelled about whether force existed in the same way as matter or was merely a property of matter, or conversely whether matter must be viewed as a product of force.

Kirchhoff did not remotely wish to answer these questions but regarded this whole manner of posing the problem as inappropriate and uninformative. However, in order to be able to refrain from any judgement about the value of such metaphysical considerations, he declared that he intended to shun all these obscure concepts completely and to confine the task of mechanics to giving the simplest and most unambiguous description of the motion of bodies without bothering about their metaphysical cause. His mechanics therefore speaks only of material points and mathematical expressions that formulate their laws of motion; the concept of force is completely absent. If Napoleon had once exclaimed in the vault of the Capuchin church in Vienna that all is vain except force, Kirchhoff banished force from nature with a single printed page, thus putting to shame the German professor who, according to Schiller's Karl Moor, was made bold to lecture on the nature of force in spite of his own weakness: for he did not annihilate it.

Kirchhoff later readopted the term force, though not as metaphysical concept but merely as an abbreviation for certain algebraic expressions that constantly occur in the description of motion. Later this word was no doubt often accorded an enhanced significance, especially in view of the analogy with muscular exertion so familiar to man, but the old obscure type of question and concept will surely never recur in natural science.

Kirchhoff made no material changes in the old classical mechanics; his reformation was purely formal. Hertz went much further; but whereas almost all later authors imitated Kirchhoff's mode of presentation, though often they merely adopted certain expressions found in Kirchhoff rather than his spirit, I have often heard Hertz's mechanics praised yet never seen anybody pursue the path he indicated.

So far as I know, nobody has yet pointed out that a certain idea in Kirchhoff's mechanics if followed to its logical conclusion leads directly to Hertz's ideas. For Kirchhoff defines the most important concept of mechanics, namely that of mass, only for the case in which there are arbitrary constraints between the material points. In that case it is easy to see that the factor Kirchhoff calls mass is necessary. In other cases where material points move without constraints in such a way as corresponded to the old effects of force, for example in elasticity, aerodynamics and so on, Kirchhoff's concept of mass floats in thin air and the resulting lack of clarity vanishes only if these cases are simply excluded.

This is what Hertz did do. The most important forces of the old mechanics were direct actions at a distance between any two material points. As to the question of the metaphysical cause of this action at a distance Kirchhoff removed it from mechanics; but motions that occur precisely according to the same laws as if forces at a distance did exist, he admitted. Now as we have seen, today we are convinced that electric and magnetic effects are transmitted through a medium. There remains only gravitation, of which Newton who discovered it already assumed that it must very likely be attributed to a medium, and molecular forces. The latter can be replaced approximately by the condition of constancy of shape in solids and of volume in liquids. However to this day it has not been possible to use conditions such as these to replace elasticity, the expansion of compressible fluids, forces of crystallization or chemical forces. Yet unlike Kirchhoff, Hertz, evidently assuming that it will be possible, rejects all motions that occur in the way required by the old forces at a distance, admitting only those motions for which there are constraints of which he gives precise mathematical definition. Beside these conditions, the only other thing he uses for constructing the whole of mechanics is a law of motion that represents a special case of Gauss's principle of least constraint.

While, then, Kirchhoff derided merely the question of the cause of motions that others attributed to forces at a distance, Hertz eradicates these motions themselves, seeking to explain the forces by means of constraints whereas the usual procedure conversely sought to explain constraints in terms of forces. It is thus in a much more real sense Hertz rather than Kirchhoff who attempts to overpower force itself. In this way he created a strikingly simple system of mechanics, starting from very few principles that present themselves automatically as it were. Unfortunately, in the same moment his lips became for ever sealed to the thousand requests for clarification that are certainly not on the tip of my tongue alone.

After what has been said, it is understandable that certain phenomena such as the free motion of rigid systems, are easily derived from Hertz's theory. For other phenomena Hertz must postulate the presence of hidden moving masses whose intervention in the motion of visible masses is required to explain the latter's laws of motion, which thus corresponds to the equally hidden medium that produces electromagnetic and gravitational effects. But how are we to conceive of these totally unknown masses from case to case? Indeed, is it possible at all in every case to attain our goal by this means? We cannot ascribe to them the structure of customary media of the past or even of Maxwell's luminous aether, since in all these media one assumed precisely those forces acting that Hertz excludes.

In even quite a simple mechanical example I was unable to find hidden masses that would lead to the goal and therefore put the problem to the Society of Natural Scientists for solution*; for so long as even in the simplest cases no systems or only unduly complicated systems of hidden masses can be found that would solve the problem in the sense of Hertz's theory, the latter is only of purely academic interest.

I therefore think that Hertz' mechanics is more a programme for the distant future. Should people one day succeed in explaining without artificiality all natural phenomena by means of hidden motions, then the old mechanics would be superseded by the Hertzian. Until then the former alone can represent all phenomena in a really clear manner without adducing things that are not only hidden but of which we have not the slightest idea how we are to conceive of them.

In his book on mechanics Hertz has given a certain completion not only to Kirchhoff's mathematico-physical ideas but also to Maxwell's epistemological ones. Maxwell had called Weber's hypothesis a real physical theory, by which he meant that its author claimed objective truth for it, whereas his own account he called mere pictures of phenomena. Following on from there, Hertz makes physicists properly aware of something philosophers had no doubt long since stated, namely that no theory can be objective, actually coinciding with nature, but rather that each

---

* *Editor's note*: Boltzmann put this question to the 70th *Versammlung der Naturforscher* etc. at Düsseldorf in 1898. See *WA* III. **129**.

theory is only a mental picture of phenomena, related to them as sign is to designatum.

From this it follows that it cannot be our task to find an absolutely correct theory but rather a picture that is, as simple as possible and that represents phenomena as accurately as possible. One might even conceive of two quite different theories both equally simple and equally congruent with phenomena, which therefore in spite of their difference are equally correct. The assertion that a given theory is the only correct one can only express our subjective conviction that there could not be another equally simple and fitting image.

Many questions that used to appear unfathomable thus fall away of themselves. How, it used to be said, can a material point which is only a mental construct, emit a force, how can points come together and furnish extension, and so on? Now we know that both material points and forces are mere mental pictures. The former cannot be identical with something extended, but can approximate as closely as we please to a picture of it. The question whether matter consists of atoms or is continuous reduces to the much clearer one, whether the continuum is able to furnish a better picture of phenomena.

We have just been speaking mainly about mechanics. The rapidly growing importance of the principle of energy has led to an attempted revolution involving the whole of physics. We mentioned this principle quite incidentally above as being a consequence, well tested by experiment, of the mechanist view of nature. According to this view, energy appears as a known mathematical expression formed in prescribed manner from previously introduced magnitudes (mass, velocity, force, path), bare of all mystery; and since this view regards heat, electricity and so on as forms of motion whose character is in part, of course, quite unknown, it sees in the energy principle an important confirmation of its conclusions.

Incidentally we meet an appreciation of this principle already in the infancy of mechanics. Leibniz spoke of the substantiality of force, by which he meant energy, almost in the same words as modern followers of energetics; but he considers that in unelastic impact *vis viva* or kinetic energy gives rise to deformations, breaks in coherence and texture, tensions of springs and so on; that heat is a form of energy he had not the slightest notion. Thus Dubois-Reymond is quite wrong as to the facts when

in his memorial speech for Helmholtz he seems once again to belittle
Robert Mayer, denying him priority in the discovery of the equivalence
of heat and mechanical work. By the way, Mayer was not at all of the
opinion that heat was molecular motion, rather he believed it to be a
totally new form of energy and maintained only its equivalence to
mechanical energy. Physicists too, who held the former view, above all
Clausius, distinguished strictly between the propositions that follow only
from it (special thermodynamics) and those that can be derived from
known facts of experience regardless of any hypothesis as to the nature
of heat (general thermodynamics).

Whereas special thermodynamics after a series of brilliant results
ground to a halt because of the difficulties in treating molecular motions
mathematically, general thermodynamics achieved a great wealth of
results. It was found that temperature decides when and how much heat
and work are transformed into each other. The gain of heat supplied was
represented as the product of the so-called absolute temperature and the
gain of another function which following Clausius we call entropy. From
the latter new functions were constructed, especially by Gibbs, such as
those that were later called thermodynamic potentials at constant temperature, constant pressure, and so on; by means of them the most surprising results were obtained in the most various fields, such as chemistry,
capillarity and so on.

Further, it was found that equations of analogous form held also for
the transformation into each other of the other forms of energy, electric,
magnetic, radiant and so on, and that again we can everywhere decompose
the expression into two factors, with similar results. This caused such enthusiasm amongst some scientists who called themselves energeticists,
that they taught that we must necessarily break with all past conceptions,
against which they urged that the inference from equivalence of heat and
kinetic energy to their identity was invalid, as though this identity was
warranted merely by the principle of equivalence and not by so much else
besides.

This new doctrine regards the concept of energy as the only valid point
of departure for enquiry into nature, and the divisibility into two factors
with a consequent variational theorem as the fundamental law of all
nature. Every mechanical model suggesting why energy takes on just
these peculiar forms in each of which it follows similar but nevertheless

significantly different laws, they regard as superfluous and even injurious; physics, and indeed the whole of future natural science, is to them a mere description of the behaviour of energy, which of course becomes pleonastic if by energy we mean anything at all that has effects.

Without doubt the analogies of the ways the different forms of energy behave are so important and interesting that pursuing their every ramification must be termed one of the noblest tasks of physics; moreover, the concept of energy is so important that it justifies the experiment of choosing it as the starting point. We must further admit that the orientation of research that I have called classical theoretical physics led to occasional excrescences against which a reaction was necessary. Every Tom, Dick and Harry felt the call to excogitate some structure, some vortices and concatenations, of atoms, believing thereby to have spotted the Creator's plan once and for all.

I know how profitable it is to attack problems from the most varied angles, and my heart beats warmly for every original and enthusiastic endeavour in science. I therefore salute the secession. I merely felt that energetics often let itself be deceived by superficial and purely formal analogies, that its laws lacked the clear and unambiguous formulation customary in classical physics, that its inferences lacked the strictness that had been perfected there, and that it rejected much of the old possessions that were good and indeed essential for science. Moreover I felt that the controversy about whether matter or energy was the truly existent constituted a relapse into the old metaphysics which people thought had been overcome, an offence against the insight that all theoretical concepts are mental pictures.

If in all these matters I have expressed my conviction without reserve, I felt that this would be a more useful proof of my interest in the further development of the theory of energy than praise could be. Just as in Hertz's mechanics, I can therefore see this doctrine, that all physics is deducible from the theorem of the ubiquitious two energy factors and the variational theorem mentioned, only as an ideal for the distant future. Only the future can tell what is so far quite undecided, namely whether such a picture of nature is better than the previous one or the best of all.

After the energeticists we come next to the phenomenologists, whom I would call moderate secessionists. Their doctrine is a reaction against the fact that the old method of research had regarded hypotheses concerning the nature of atoms as the proper goal of science, while viewing the resulting laws for visible processes as rather merely means for testing the hypotheses.

This indeed holds only for the most extreme orientation within the old method. We have seen that already Clausius had strictly distinguished between general thermodynamics, which is independent of molecular hypotheses, and special thermodynamics. Many other physicists too, for example Ampère, Franz Neumann, Kirchhoff, did not base their derivations on molecular ideas, even if they did not deny the atomistic structure of matter.

Here we often find a mode of derivation that I shall call Euclidean, since it is fashioned after the procedure used by Euclid in geometry. Certain propositions (axioms) are presupposed either as self-evident or at least as fixed beyond doubt by experience; from these one next derives certain simple elementary theorems as logical consequences and only then we construct from these last the general integral laws.

With this method and those based on molecular theory it had hitherto been more or less possible to make do; not so in the case of Maxwell's theory of electro-magnetism. In his first writings Maxwell conceived of the medium that propagated electro-magnetism as consisting of a large number of molecules or at least of mechanical individuals, but of so complicated a structure that they could be valid only as an auxiliary device for finding equations and as schemata for effects that resembled actual ones in certain respects, but never as definitive pictures of what existed in nature. Later he showed that these were not the only mechanisms that led to his goal, but that many others would do so too, so soon as they satisfied certain general conditions; but all efforts at finding a definite and really simple mechanism satisfying all these conditions proved unsuccessful. This paved the way for a doctrine that seems describable in the most suggestive manner if for a third time we go back to Hertz, whose ideas set down in the introduction to his essay on the fundamental equations of electro-dynamics are typical of this doctrine.

Hertz was not looking for a satisfactory mechanical explanation of these fundamental equations, at least he did not find one; but he even spurned the Euclidean mode of derivation. He rightly points out that what convinces us of the correctness of all these equations is not, in mechanics, the few experiments from which its fundamental equations are usually derived, nor, in electrodynamics, the five or six basic experiments of Ampère, but rather their subsequent agreement with almost all hitherto known facts. He therefore passes a judgment of Solomon that since we have these equations we had best write them down without derivation, compare them with phenomena and regard constant agreement between the two as the best proof that the equations are correct.

The view whose most extreme form has here been stated, was very variously received. Whereas some were almost inclined to regard it as a bad joke, others felt that physics must henceforth pursue the sole aim of writing down for each series of phenomena, without any hypothesis, model or mechanical explanation, equations from which the course of the phenomena can be quantitatively determined; so that the sole task of physics consisted in using trial and error to find the simplest equations that satisfied certain required formal conditions of isotropy and so on, and then to compare them with experience. This is the most extreme form of phenomenology, which I should like to call mathematical, whereas general phenomenology seeks to describe every group of facts by enumeration and by an account of the natural history of all phenomena that belong to that area, without restriction as to means employed except that it renounces any uniform conception of nature, any mechanical explanation or other rational foundation. This latter view is characterized by Mach's dictum that electricity is nothing but the sum of all experience that we have had in this field and still hope to have. Both views set themselves the task of representing phenomena without going beyond experience.

Mathematical phenomenology at first fulfils a practical need. The hypotheses through which the equations had been obtained proved to be uncertain and prone to change, but the equations themselves, if tested in sufficiently many cases, were fixed at least within certain limits of accuracy; beyond these limits they did of course need further elaboration and refinement. Thus practical use alone requires us to distinguish as neatly as possible the fixed and certain parts from the changeable ones.

Besides we must admit that the purpose of all science and thus of physics too, would be attained most perfectly if one had found formulae by means of which the phenomena to be expected could be unambiguously, reliably and completely calculated beforehand in every special instance; however this is just as much an unrealisable ideal as the knowledge of the law of action and the initial states of all atoms.

Phenomenology believed that it could represent nature without in any way going beyond experience, but I think this is an illusion. No equation represents any processes with absolute accuracy, but always idealizes them, emphasizing common features and neglecting what is different and thus going beyond experience. That this is necessary if we are to have any ideas at all that allow us to predict something in the future, follows from the nature of the intellectual process itself, consisting as it does in adding something to experience and creating a mental picture that is not experience and therefore can represent many experiences.

Only half of our experience is ever experience, as Goethe says. The more boldly one goes beyond experience, the more general the overview one can win, the more surprising the facts one can discover but the more easily too one can fall into error. Phenomenology therefore ought not to boast that it does not go beyond experience, but merely warn against doing so to excess.

Even if it imagines that it is not positing a picture for nature, phenomenology is in error. Numbers, their relations and groupings are just as much pictures of processes as the geometrical ideas of mechanics. The former are merely more prosaic and more suitable for quantitative representation but for this very reason less apt to point to essentially new perspectives, they are bad heuristic road signs; likewise all ideas of general phenomenology show themselves to be pictures of phenomena. Thus the best result will no doubt be achieved if we always use all means of picturing as required, without neglecting to test the pictures at each step against new experience.

Besides, that way one will not overlook facts because one is blinded by the pictures, as has been laid to the charge of atomists. Any theory of whatever kind will lead to this form of blindness if it is pursued too one-sidedly. It was less the fault of some specific peculiarity of atomism than of the fact that people were as yet insufficiently warned against placing too much confidence in the pictures. No more must mathematicians confuse their formulae with truth, or they will be similarly blinded. This is seen with phenomenologists, when they fail to notice the many facts that can be grasped only from the point of view of special thermodynamics, with opponents of atomism when they ignore everything that supports the doctrine and even with Kirchhoff when confidently basing himself on his hydrodynamic equations he regards it as impossible for the pressure to differ at different points of a heat-conducting gas.

Mathematical phenomenology, as lies in its nature, went back to the idea of continuity of matter, which corresponds to appearances. As against this I have pointed out that the differential equations by definition represent only transitions to the limit, which without supposing the idea of a very large number of individual beings would simply be senseless. Only if one uses mathematical symbols thoughtlessly can one imagine that differential equations can be severed from atomistic ideas. If we become fully aware that phenomenologists also start from atom-like individual beings, disguised in the clothes of differential equations, although they must conceive of them as different for each group of phenomena, being endowed now with this now with that property in the most complicated manner, then the need for a simplified uniform atomism will soon reassert itself.

Energeticists and phenomenologists had inferred from the small current yield of molecular theory to its decline. While according to some this theory had in any case been only harmful, others nevertheless admitted that it was of some use in the past and that almost all equations that to the phenomenologists now count as the very essence of physics had been obtained by way of molecular theory, but that it had become superfluous now that we already have these equations. They all swore to annihilate it. They pointed to the historical principle that often the views held in highest esteem are soon displaced by quite different ones; indeed as St Remigius had advocated the burning of heathen, so these theoretical physicists advocated the burning of what had still been venerated only a moment ago.

However, historical principles are at times double-edged. Certainly history often shows unforeseen revolutions; certainly it is useful to keep in mind the possibility that what now seems to be most secure may one day be displaced by something quite different; but equally the possibility that certain achievements will remain permanently established in science even if in amplified and altered form. Indeed, according to the historical principle mentioned energeticists and phenomenologist could never be definitively victorious, for that would imply at once that their renewed downfall was imminent.

Following Clausius, the adherents of special thermodynamics have never denied the great value of general thermodynamics, whose success thus proves nothing against the special theory. We may enquire only whether alongside these results there are also others that could be obtained only through atomism, and of such results atomism still affords many, even long after its old period of pre-eminence. Purely from principles of molecular theory, Van der Waals has derived a formula that describes the behaviour of fluids, gases and vapours and the various forms of transition of these states of aggregation, not indeed with complete accuracy but with admirable approximation, while leading to many new results, for example the theory of corresponding states. Quite recently, considerations of molecular theory have shown further ways of improving the formula, and we need not exclude the hope that we may be able to represent with perfect accuracy the behaviour of the chemically simplest substances, particularly argon, helium and so on; thus it is precisely atomism that has most nearly approached the phenomenologists' ideal of a mathematical formula that covered all states of a body. Following on from this there was a kinetic theory of liquids.

Moreover, atomism has in recent times contributed much to giving a model for and elaboration of Gibbs' theory of dissociation, which he had discovered by a different method though still presupposing certain basic conceptions of molecular theory. Atomism not only gave a new foundation of the equations of hydro-dynamics but also showed where they and indeed the equations of thermal conduction still need correcting. No doubt phenomenology too regards it as desirable always to conduct new experiments in order to find corrections that its equations may need, still, atomism is here much more efficient in that it enables us to point to definite experiments that most directly offer prospects of actually finding such corrections.

The distinctively molecular theory of the ratio of the two specific heats of gases has likewise resumed an important role today. For the simplest gases, whose molecules behave like elastic spheres, Clausius had worked out the value of this ratio at $1\frac{2}{3}$, which did not fit any gas then known; from which he concluded that there are no gases of such simple constitution. For the case in which the molecules behave like non-spherical elastic bodies, Maxwell found the value $1\frac{1}{3}$. Since for the best known gases the ratio was 1.4, Maxwell too rejected his theory. But he had overlooked the case in which the molecules are symmetrical about one axis, in which case theory requires precisely 1.4 as the value for the ratio in question.

Already Kundt and Warburg had found the old Clausius value of $1\frac{2}{3}$ for mercury vapour, but because the experiment was difficult it had never been repeated and all but forgotten. Then the same value turned up again for the ratio of the specific heats in all gases newly discovered by Lord Rayleigh and Ramsay, and all other circumstances pointed to the specially simple molecular structure required by theory, as had been the case already for mercury vapour. What influence it might have had on the history of gas theory, if Maxwell had not made this slight mistake, or if the new gases had been known as early as the time of Clausius! From the very start one would have found in the simplest gases the value that theory required for the ratio of the specific heats.

Let me finally mention the relation that molecular theory points out between the entropy theorem and the calculus of probability, whose real meaning may indeed be controversial but of which no unprejudiced person will surely deny that it is capable of widening the horizon of our ideas and giving us hints towards new combinations of concepts or even experiments.

All these achievements and many earlier attainments of atomic theory are absolutely unattainable by phenomenology or energetics; and I assert that a theory that achieves original insights unobtainable by other means, that is moreover supported by many facts of physics, chemistry and crystallography, such a theory should not be opposed but cultivated further. As regards ideas about the nature of molecules it will however be necessary to leave the widest possible room for manoeuvre. Thus one will not give up the theory of the ratio of specific heats merely because it is not yet generally applicable; for molecules do not behave like elastic bodies except for the simplest gases and even for them not at very high temperatures and only as to their collisions; about their more detailed constitution which is bound to be very complicated we have as yet no indications, rather we shall have to try to find them. Parallel to atomism the further precision and discussion of the equally indispensable equations can proceed divorced from any hypothesis, without the former raising to the status of dogma its material points, or the latter its mathematical apparatus.

Until now, however, the liveliest controversy of views goes on; each thinks his own is genuine, and well he may, if it is done with the intention of testing its power as against the others. Rapid progress has stretched our expectations to the limit; what will be the end?

Will the old mechanics with its old forces, even if divested of metaphysics, continue to exist in its basic features or one day merely live on in history, displaced by Hertz's hidden masses or by some quite different ideas? Of present day molecular theory, notwithstanding any additions and modifications, will the essential features nevertheless survive, or will the future one day bring an atomic theory that is totally different from today's; or, contrary to my demonstration, will it be found one day that the idea of a pure continuum affords the best picture? Will the mechanist view of nature one day win the decisive battle for the discovery of a simple mechanical picture of the luminous aether, will mechanical models at least always continue to exist, will new and non-mechanical ones prove to be superior, will the two factors of energy one day rule everything, or will people in the end be content to describe every agent as the sum of all kinds of phenomena, or will theory turn into a mere collection of formulae and the attendant discussion of equations?

More generally, shall we ever come to be convinced that certain pictures can no longer be displaced by simpler and more comprehensive ones, that they are 'true', or do we perhaps obtain the best idea of the future if we imagine that of which we have no idea at all?

Interesting questions indeed! One almost regrets having to die long before they are decided. How immoderate we mortals are! Delight in watching the fluctuations of the contest is our true lot.

For the rest, it is better to work on what lies close to hand rather than rack one's brain about such remote questions. Our century has achieved much indeed! To the next, it bequeathes an unexpected wealth of positive facts and a splendid sifting and purification of methods of research. A Spartan chorus of warriors called out to the young men: Grow up to be yet braver than we! If, following an old custom we wish to greet the new century with a blessing, we can truly wish, equalling those Spartans in pride, that it may come to be even greater and more significant than the departing one!

ON THE FUNDAMENTAL PRINCIPLES AND
EQUATIONS OF MECHANICS*

I

Analytical mechanics is a science worked out by its very founder Newton with a precision and perfection almost unrivalled in the whole field of human knowledge. The great masters that succeeded him have further strengthened the structure he had erected, so that it seemed quite inconceivable that there could be a creation of the human spirit more perfect and uniform than the foundations of mechanics as they confront us in the works of Lagrange, Laplace, Poisson, Hamilton and so on. Especially the establishment of the first principles seemed to have been carried out by these enquirers with a precision and logical consistency that has always furnished the paradigm according to which people sought to fashion the foundations of the other branches of knowledge, if not always with the same success. It long seemed quite impossible to expand or modify those foundations in any way.

It is the more noticeable and unexpected that at present there have arisen, especially in Germany, fairly lively controversies precisely about the fundamental principles of analytical mechanics. This must of course not be taken to mean that the awe and admiration we accord to geniuses like Newton, Lagrange or Laplace is in any way to be diminished, for from the small beginnings that they found they have created a masterly paradigm for all time. They had to elicit so much that was actually new that it would merely have caused delay and damage to the unitary impression if they had tarried too long with certain difficulties and obscurities. However, since then our factual knowledge has significantly grown and our intelligence is schooled, so that many ideas that in Newton's time were still causing difficulties to scholars have now become the common property of all. This provided the leisure for looking at the construction of the Newtonian edifice through a magnifying glass as it were, and lo and

* *Populäre Schriften*, Essay 16. The first two of four lectures given at Clark University, 1899.

behold, this yielded many difficulties such as always tend to confront the human spirit precisely where it strives after an analysis of the simplest foundations of knowledge.

To be sure, these difficulties are more of a philosophical nature, or epistemological as the current jargon has it. We Germans have often been much laughed at for our leaning towards philosophic speculation, and in earlier times no doubt often justly so. A philosophy that turns away from fact has never produced anything useful nor can it do so. It is above all of direct and tangible use to widen our factual knowledge by means of experiments and our knowledge of nature too is primarily and most productively enhanced in this way. But in spite of all this there seems to be an invincible streak in the human spirit, prompting it to analyse the simplest concepts and to render an account of the fundamental operations of our own thinking.

In the course of time this method of analysis was also considerably perfected, so that today it is no longer nearly so empty as the old philosophy, even if by no means yet of immediate practical use. Indeed, in the course of history the whole cultural aspect of mankind constantly suffers considerable fluctuations. Germans no longer are the unpractical dreamers of yesterday, as they have shown in all fields of experimental science, engineering, industry and politics. The endeavours of Americans were naturally at first directed towards the practical activities of industry and engineering, for the sake of mastering the country itself. But they no longer are exclusively that way inclined, for in all fields of abstract science America already has enquirers who are fully equal to the most eminent Europeans. Since therefore you have invited a German to lecture in your country, I will venture to proceed with you into an area of epistemology.

Let me first return to the reservations that have been urged against the foundations of Newtonian mechanics, or better, to those points where they still seem to need closer elucidation and an analysis of the modes of inference and sifting of concepts. In formulating the laws of motions Newton regards the motion of bodies as being absolute in space. However, absolute space is nowhere accessible to our experience, which only ever gives us the relative changes of position of bodies. Thus from the outset this goes completely beyond experience, which must surely be questionable in a science that sets itself the task of representing facts of experience. This difficulty did of course not escape the genius of Newton. However, he thought that without the concept of an absolute space he could not reach a simple formulation of a law of inertia which was his first concern and I believe that he was right in this; for however much this difficulty was elucidated or thought through, there has been hardly any progress. Instead of Newton's absolute space, Neumann introduced a mysterious ideal body of reference, which as with Newton evidently goes beyond experience. Streintz sets himself the task of avoiding such concepts or bodies, by showing how using the motion, relative to a chosen co-ordinate system, of a gyroscope subject to known forces or none, we can decide whether for this system Newton's laws of motion hold so that it is a useful reference system. However these considerations of Streintz's seem of little use for the foundations of mechanics, because they already presuppose the laws of motion of a spinning top and a judgment as to the action of forces on it or otherwise, which in turn already requires knowledge of Newton's laws of motion. Lange indeed tries to formulate the law of inertia without any reference system, merely by considering relative motion and he even succeeds, but his account is so complicated and prolix that one will be reluctant to put so unperspicuous a law in place of the simple Newtonian formula. Obviously Mach's suggestion of straight lines determined by the totality of all masses in the world or the suggestion of adopting the luminous aether in place of absolute space both go beyond experience although of course in quite different ways. For the former starts once again from purely ideal and transcendental concepts whereas the latter makes an assertion that might possibly be confirmed by experience but certainly is not so at present. The aether would have to obey quite a different mechanics, for it would have to be the cause of inertia but not be subject to it. We meet similar difficulties when we introduce the concept of time. This too is introduced by Newton as absolute, whereas this is never given to us: we always merely experience simultaneity of the course of several processes. However the remedy is easier here, for we can start from a process that always repeats itself periodically under exactly similar circumstances. Of course it is impossible to produce absolutely identical circumstances, but we can make it eminently likely that all circumstances of any essential bearing are the same. We can further confirm this by mutually comparing different kinds of processes that have this property (rotation of the Earth, oscillations of a pendulum or of a chronometer spring). The agreement of all these processes in indicating the same time then excludes all doubt as to the usefulness of this method.

A third difficulty concerns the concepts of mass and force. That Newton's definition of mass as quantity of matter says nothing has long been recognized. However, doubts arise also as to the ratio of force to mass. Does mass alone exist, force being merely a property of it, or conversely, is force alone truly existent, or must we assume a dualism of two separate existents (mass and force), force existing independently of matter and causing its motion. In addition there is the more recent question whether energy too must be accorded existence or whether indeed it alone exists.

It was above all Kirchhoff who on this point objected to the very question itself. Often a problem is half solved as soon as the right way of asking the question has been found. Kirchhoff rejected the notion that it was the task of science to unravel the true nature of phenomena and to state their first and fundamental metaphysical causes. On the contrary he confined the task of natural science to describing phenomena, a stipulation that he still called a restriction. If however we delve really deeply into the mode of our own thinking, into its mechanism, as I should feel inclined to say, then one should like to deny even that proviso.

All our ideas and concepts are only internal mental pictures, or if spoken, combinations of sounds. The task of our thinking is so to use and combine them that by their means we always most readily hit upon the correct actions and guide others likewise. In this, metaphysics follows the most down-to-earth and practical point of view, so that extremes meet. The conceptual signs that we form thus exist only within us, we cannot measure external phenomena by the standard of our ideas. We can therefore pose such formal questions as whether only matter exists and force is a property of it, or whether force exists independently of matter or conversely whether matter is a product of force; but none of these questions are significant since all these concepts are only mental pictures whose purpose is to represent phenomena correctly. This was stated with special clarity by Hertz in his famous book on the principles of mechanics, except that he there begins with the demand that the pictures we construct for ourselves must obey the laws of thought. Against this I should like to urge certain reservations or at least to explain the demand a little further. Certainly we must contribute an ample store of laws of thought, without them experience would be quite useless, since we could not fix it by means of internal pictures. These laws of thought are almost without exception innate, but nevertheless they suffer modification through upbringing, education and our own experience. They are not quite the same in a child, a simple uneducated person, or a scholar. We can see this too when we compare the mental orientation of a naive people like the Greeks with that of the mediaeval scholastics and theirs in turn with that of today. Certainly there are laws of thought that have proved so sound that we place unconditional confidence in them, regarding them as unalterable a priori principles of thought. However, I think nevertheless that they developed gradually. Their first source was the primitive experience of mankind in its primeval state, and gradually they grew stronger and clearer through complicated experience until finally they assumed their precise present formulation; but I do not wish to recognize the laws of thought as supreme arbiters. We cannot know whether they might not suffer this or that modification in future. Let us remember how certain children or uneducated people are in their conviction that our feeling alone must be able to decide what is up and down at all points of space from which they imagine they can deduce that antipodes are impossible. If such people wrote on logic, they would very likely regard this as an a priori self-evident law of thought. Just so people raised many a priori objections to the Copernican theory and the history of science contains many cases where propositions were either founded on or refuted by arguments that at the time were regarded as self-evident laws of thought, whereas we are now convinced that they are futile. I therefore wish to modify Hertz's demand and say that insofar as we possess laws of thought that we have recognized as indubitably correct through constant confirmation by experience, we can start by testing the correctness of our pictures against these laws; but the sole and final decision as to whether the pictures are appropriate lies in the circumstance that they represent experience simply and appropriately throughout so that this in turn provides precisely the test for the correctness of those laws. If in this way we have grasped the task of thought in general and of science in particular, we obtain conclusions that are at first sight striking. We shall call an idea about nature false if it misrepresents certain facts or if there are obviously simpler ideas that represent these facts more clearly and especially if the idea contradicts generally confirmed laws of thought; however, it is still possible to have theories that correctly represent a large number of facts but are incorrect in other aspects, so that they have a certain relative truth. Indeed, it is even possible that we can construct a system of pictures of experience in several different ways. These systems are not all equally simple, nor represent phenomena equally well. However it may be doubtful and in a sense a matter of taste which representation satisfies us most. By this circumstance science loses its stamp of uniformity. We used to cling to the notion that there could be only one truth, that error was manifold but truth one. From our present position we must object to this view, although the difference between the old and new views is more of a formal kind. It was never in doubt that man would never be able to know the full essence of all truth, such knowledge is only an ideal. However, according to our present view too, we possess a similar ideal, namely that of the most perfect picture representing all phenomena in the simplest and most appropriate manner. Thus, according to the one view we turn our gaze more towards the one unattainable unitary ideal, according to the other towards the multiplicity of what is attainable.

If now we are convinced that science is only an internal picture or mental construction that can never coincide with the multiplicity of phenomena but only represent certain parts of them in an ordered manner, how are we to reach such a picture and how to represent it with all due order and system? At one time the method that was popular was one fashioned after Euclid's, used in geometry, which we may thus call Euclidean. This method starts from as few and obvious propositions as possible. At first these were regarded as a priori self-evident and directly given to the mind, and were therefore called axioms. Later one ascribed to them the character of merely sufficiently well guaranteed propositions of experience. From these axioms and with the help of the laws of thought one then simply deduced certain pictures as necessary, believing thus to have found a proof that they were the only possible ones and not replaceable by any others. As example take the arguments that served for deriving the parallelogram of forces or Ampère's law or for proving that the force between two material points acts along the line joining them and is a function of their distance.

However, the force of this argument became gradually discredited, the first step towards this being the earlier mentioned transition from an a priori self-evident foundation to one that was merely established by experience. Further, it was realised that the deductions from that foundation could not be made either except with the help of many new hypotheses, so that Hertz finally pointed out that especially in the field of physics our conviction of the correctness of a general theory does not rest essentially on its deduction according to the Euclidean method but rather on its leading in all cases hitherto known to correct inferences as regards phenomena. He first used this view in his account of Maxwell's fundamental equations in the theory of electricity and magnetism, where he proposes not to trouble himself about deducing them from given fundamental principles, but simply to put them at the beginning and to seek a justification of this procedure in our being able to show their subsequent agreement in every point with experience, for in the end the only judge as to whether a theory is usable remains experience, and its verdict cannot be shaken or appealed against. Indeed, if we take a closer look at the theorems most intimately connected with the topic, namely the law of inertia, the parallelogram of forces and the other fundamental theorems of mechanics, we shall find the different proofs given for them in various treatises on mechanics not nearly so convincing as the fact that all consequences derived from this complex of theorems have been so excellently confirmed by experience. The ways in which we reached the pictures are not infrequently most disparate, and contingent on the most varied accidents.

Some pictures were built up only gradually over centuries through the joint efforts of many enquirers, for example the mechanical theory of heat. Some were found by a single scientific genius, though often by very intricate detours, only then could other scientists illuminate them from various angles. Maxwell's theory of electricity and magnetism discussed above is one such. Now there is no doubt a particular mode of representation that has quite peculiar advantages, though it has its defects too. This mode consists in starting to operate only with mental abstractions, in tune with our task of constructing only internal mental pictures. In this we do not yet take account of facts of experience. We merely endeavour to develop our mental pictures as clearly as possible and to draw from them all possible consequences. Only later, after complete exposition of the picture, do we test its agreement with the facts of experience; it is, then, only after the event that we give reasons why the picture had to be chosen thus and not otherwise, a matter on which we give not the slightest prior hint. Let us call this deductive representation. Its advantages are obvious. For a start, it forestalls any doubt that it aims at furnishing not things in themselves but only an internal mental picture, its endeavours being confined to fashioning this picture into an apt designation of phenomena. Since the deductive method does not constantly mix external experience forced on us with internal pictures arbitrarily chosen by us, this is much the easiest way of developing these pictures clearly and consistently. For it is one of the most important requirements that the pictures be perfectly clear, that we should never be at a loss how to fashion them in any given case and that the results should always be derivable in an unambiguous and indubitable manner. It is precisely this clarity that suffers if we bring in experience too early, and it is best preserved if we use the deductive mode of representation. On the other hand, this method highlights the arbitrary nature of the pictures, since we start with quite arbitrary mental constructions whose necessity is not given in advance but justified only afterwards. There is not the slightest proof that one might not excogitate other pictures equally congruent with experience. This seems to be a mistake but is perhaps an advantage at least for those who hold the above-mentioned view as to the essence of any theory. However, it is a genuine mistake of the deductive method that it leaves invisible the path on which the picture in question was reached. Still, in the theory of science especially it is the rule that the structure of the arguments becomes most obvious if as far as possible they are given in their natural order irrespective of the often tortuous path by which they were found. In his book on mechanics Hertz has given a sample of such a purely deductive representation from that field too. I presume I may take the contents of the book as known so that a very brief account should here suffice. Hertz starts from material points regarded as purely mental pictures. Mass, too, he defines quite independently of all experience by means of a number to be thought of as associated with each material point, namely the number of simple mass points that it contains. From these abstract concepts he constructs motion which like the points themselves is of course a mental construction. The concept of force is completely absent and in its stead we have certain constraining conditions expressed in terms of the differentials of the coordinates belonging to the material points. These latter are now furnished with given initial velocities and thereafter always move according to a very simple law which unambiguously fixes the motion for all time as soon as the constraints are given. Hertz formulates it thus: the sum of the squared deviations, each multiplied by the mass, of material points from uniform rectilinear motion at each moment must be a minimum, or more succinctly still, the motion follows the straightest paths. This law is very similar to Gauss's principle of least constraint, indeed it is that special case that supervenes when the principle is applied to a system of points subject to a constraint but to no other external forces.

In my *Lectures on the Principles of Mechanics*, I too have attempted a purely deductive account of the fundamental principles of the subject, but in quite a different way, much more related to the conventional method of treatment. Like Hertz, I start from purely mental constructs, namely exact material points, whose position is given in an imagined rectangular co-ordinate system. My mental picture of their motion is then at first as follows: whenever two points are at a distance r from each other, each is accelerated in the direction of r by an amount that is a function $f(r)$ to be specified later as required. Moreover, the ratio of the two accelerations should be permanently fixed, so defining the ratio of the two masses. As to the motion of all material points together, that is to be considered as uniquely determined by the stipulation that the actual acceleration of each point is the vector-sum of all of its accelerations according to the rule just stated. The effect of these accelerations is added vectorially to the speed already acquired, but there is no discussion as to where they come from and why I have given just this prescription for constructing our picture in just this way. It is enough that the picture is perfectly clear and in sufficiently many cases capable of being worked out in detail by calculation. Its justification is found only in the fact that we can always define $f(r)$ in such a way that the imagined motion of the imagined material points goes over into a true representation of actual phenomena.

In using this, which we have called the purely deductive, method we have of course not solved the question as to the nature of matter, mass and force, but we have circumvented it by making a prior consideration of it quite superfluous. In our conceptual scheme these concepts are quite definite numbers and directions for geometrical constructions which we know how to consider and execute in order to obtain a useful picture of the phenomenal world. What might be the true cause for this world to run as it does, what is as it were concealed behind it and acts as its motor, these things we do not regard as the business of natural science to explore. Whether it might or could be the task of some other science, or whether following the analogy of other, sensible collocations of words we have perhaps merely strung together words that express no clear thought in these combinations, all this may be left entirely open here. Just as little have we solved the question of absolute space and absolute motion by this deductive method; however, this question too no longer involves pedagogical difficulties, we no longer need to mention it when beginning to develop the laws of mechanics, but can defer its discussion until all these laws have been derived. For since at the start we only introduce mental constructions in any case, an imagined co-ordinate system does not look out of place amongst them. It is simply one of the means of construction that are intelligible and familiar to us and we use it to put together our mental picture; it is no more and no less abstract than the material points whose relative motion with regard to the co-ordinate system we imagine and for which alone at first we pronounce the laws and formulate them mathematically. On comparison with experience we then find that a co-ordinate system immutably linked with the sphere of fixed stars is practically quite adequate to ensure agreement with experience. What co-ordinate system we may have to base ourselves on one day, when we shall be able to express the motion of fixed stars by mechanical formulae, this question comes absolutely last on our plan of enquiry, and we can now easily discuss all the hypotheses of Streintz, Mach, Lange and so on which we mentioned at the beginning, since we already dispose of all the laws of mechanics. We are not embarrassed as previously, where we should have had to put these complicated considerations before the development of the law of inertia. But then the deductive method will require a proof that was superfluous with the old methods. Since with the latter we started directly from phenomena, it went without saying that the laws of phenomena cannot depend on the choice of the mentally superimposed co-ordinate system, and it was bound to strike us as odd that these laws become different and much more complicated if we introduce a rotating co-ordinate system. With the deductive method, however, we have from the outset assigned to the co-ordinate system the same role in the picture as to the material points. It is an integral part of the picture and we can hardly wonder that the picture looks different for different co-ordinate systems. On the contrary, here we must introduce arbitrarily different co-ordinate systems, so long as these have no mutual relative rotations or accelerations.

Let us now compare this method of representation of my book with that of Hertz. Classen has called my account a polemic against Hertz, presenting the case as if I imagined I had produced something unconditionally superior to Hertz. Nothing could be further from the mark. I absolutely recognise the advantages of Hertz's picture, but on the principle that it is possible and desirable to set up several pictures for one and the same group of phenomena, I think that alongside Hertz's picture mine still has its significance, in that it has certain advantages lacking in his. The principles of mechanics that Hertz sets up are extraodinarily simple and beautiful. They are naturally not entirely free from an arbitrary streak, but this is kept to a minimum. The picture that Hertz thus constructs independent of experience has a certain inner perfection and obviousness and contains really only few arbitrary elements. As against this my picture is evidently inferior, for it contains many more features that are marked by an absence of inner necessity, being introduced only in order to facilitate subsequent agreement with experience. Besides it contains a quite arbitrary function, and of the many pictures that result when this function assumes all possible forms, only a very few correspond to actual processes. Whereas with Hertz's picture we see at once that if any other pictures are possible at all then certainly only very few that possess the same simplicity and perfection, my picture immediately provokes the idea that there must be a good many other such that represent phenomena equally well. Nevertheless there are some points in which my picture is superior to Hertz's. He can indeed explain some phenomena directly from his picture, or perhaps better, represent them by means of it, such as the motion of a material point on a prescribed surface or curve, or the rotation of a rigid body about a fixed point, provided of course that there are no strange external forces. However, difficulties arise as soon as one wants to represent the most ordinary processes of daily experience involving the action of forces. Take first the most general and important forces in nature, namely gravitation. On Hertz's view we must of course not regard it as acting at a distance. Many experiments have been conducted to explain it mechanically through the action of a medium, but as is well known, none has led to a really determinate and decisive result. One of the best-known attempts is the theory of molecular impacts first put forward by Lesage and later taken up again by Lord Kelvin, Isenkrahe and others. Quite apart from the fact that it remains doubtful whether this theory can be carried through exactly, it is of no use in a Hertzian context since there, as we shall soon see, the explanation of a single elastic impact is enough to create difficulties. Thus one would first have to develop a completely new theory, explaining gravitation perhaps by means of vortices, pulsations or the like, where the particles of the medium in question likewise must not be linked by forces in the old sense but only by constraints of the type set up by Hertz. Even should this succeed, it would still amount to no more than quite an arbitrary picture which in the course of time would no doubt have to be replaced by quite a different one. Hertz's objection to classical mechanics, that it gives much too wide a picture, since of all possible functions $f(r)$ representing force only a very few have any practical use, can be turned in even greater measure against his own picture, as soon as one wishes to apply it to concrete cases. Already with gravitation we must select from all possible media that might transmit action at a distance one definite form, surely an even more indefinite and arbitrary procedure than the choice of certain functions $f(r)$.

As is well known, in his early papers Maxwell gave a successful account of electric and magnetic forces in terms of the action of a medium. Quite apart from the fact that this medium has a very complex structure and bristles with properties that bear the stamp of arbitrariness and of a purely provisional character, it would in any case be useless for Hertz, because for him too the parts are held together by forces in the traditional sense of mechanics. Indeed, even the properties of elastic, liquid and gaseous bodies would have to be replaced by new pictures, since our present ones are all based on the assumption of forces acting between particles. We are thus restricted to a simple choice: either the nature of the mechanism causing gravitation and electric and magnetic phenomena is left indeterminate and arbitrary, which produces an intolerable obscurity because one constantly has to operate with equations of which only a few quite general properties are known while their special form is quite unknown; or one endeavours to choose a definite mechanism, which again involves one in just as many arbitrary steps as difficulties.

However, let me illustrate in a much simpler example what difficulties are met in applying Hertz's fundamental law in even the most trivial cases.

Let there be three masses m_1, u, m_2 with the condition that the distance of u from either of the others is always constant and equal to a. Now let u diminish indefinitely; this case is perfectly in line with the spirit of Hertzian mechanics and gives us a true picture of the following process in nature.

In a hollow elastic sphere of mass m_2 a small solid elastic sphere is moving about. Let the difference between the two radii be $2a$. We thus have an example of one and the same natural process explicable in two quite different ways, either from molecular theory or according to the method indicated by Hertz. However, not all processes behave like this. Already the quite trivial case of elastic impact between two solid spheres can be derived from Hertz's schema only with the help of rather arbitrarily chosen mechanisms or complicated assumptions about an intervening medium, since the Hertzian method excludes inequalities. Thus even in the simplest cases, Hertz's method leads to the greatest complications.

Let me stress again that these remarks are not at all intended to deny the eminent value of the Hertzian picture, consisting in the logical simplicity of its fundamental principles. Indeed, it is conceivable that in the remote future we may one day be able to explain all action by means of media, whose properties are not chosen fancifully but imposed on us by the nature of things in an immediate and unambiguous way. It might be that the particles of these media did not exert forces on each other in the traditional sense of mechanics, but that it would be sufficient to work with Hertzian constraints between the co-ordinates of the elementary particles. From that moment, Hertz's mechanics would have won a clear victory and all other representations would retain only historical interest. Whether one considers it likely that such a point in time will supervene or not is of course purely a matter of taste. We have not even proved that such a development of our knowledge is possible. From our present point of view we shall therefore look up to this ideal with admiration and make our own contribution towards promoting an approach to it. For the time being, however, alongside Hertzian ones we shall not be able to forgo simple and directly useful pictures that can be worked out in detail.

II

In my previous lecture I discussed two pictures of mechanical phenomena, namely Hertz's and my own as given in my book on mechanics. Mine is not essentially different from older theories of mechanics. I merely tried, by means of as consistent an account as possible, to secure mechanics against any objections and in particular against the reservations that Hertz makes with regard to the older mechanics in the preface of his book.

Precisely for this purpose a purely deductive account seemed the most suitable, because it allows us to develop a picture quite independently of the facts, and that with the utmost clarity. However, one could develop the picture in the opposite direction too, by starting directly from the facts as they present themselves to unprejudiced observation, letting the pictures grow gradually from these facts and introducing each abstraction only when there is no way left of avoiding it. This we shall call the inductive mode of presentation, which compared with the deductive has the drawback that the pictures do not from the outset stand in their purest form, so that their internal consistency cannot be so clearly surveyed. On the other hand it has the advantage that instead of the abstract presentation of the deductive method with its back turned to reality*, it favours a method that starts purely from the immediately given and familiar, showing up as clearly as possible how the abstract pictures arose and why we resort to precisely those pictures. To compare the advantages and drawbacks of the two methods it would be not inappropriate to compare the method outlined in the previous lecture with the older modes of representation usual in mechanics since these last intermingle the two methods, which in my view impairs clarity. Thus as a rule abstract concepts like material points, mass and so on are introduced fairly early but without being considered as purely conceptual tools, as we did in the last lecture. Rather, we are given more or less indefinite and empty definitions of them. For example, a material point is defined as a body that is so small that its extension can be neglected. This is meant to convey perhaps that its moments of inertia about an axis through its centre of gravity vanish in comparison with those about some other axis at a distance of the order common in ordinary experiments, or the like. But since the concept of moment of inertia, centre of gravity and so on have not yet been developed, I should not know what to understand by a body one of whose most important properties, namely extension, can be neglected. Mass is often defined by the action of one and the same force on different bodies, but how is one to observe that the force is the same if it acts now on this body now on that? It will thus be best if we try to sketch yet another purely inductive method for presenting the basic principles of mechanics at least in outline. In this we remain faithful to our principle that for the time being

---

* Translator's note: Reading '*Wirklichkeit*' for the implausible '*Willkürlichkeit*' (arbitrariness).

we are not aiming at a single best account of science, but that we regard it as expedient to try as many accounts as possible, each of which has its peculiar advantages but also its drawbacks. Again we must focus our main attention on avoiding all inconsistencies and logical mistakes and on not smuggling in tacit concepts or assumptions, and ensure on the contrary that we become most clearly aware of all hypotheses we rely on. It goes without saying that I cannot here, in the short time at my disposal, give an exhaustive account of mechanics. I shall merely try to give a few hints. Nor, indeed, would it be possible to solve so difficult a problem at one stride. Much will remain deficient in the first attempt and only gradually will the concepts be sifted and the account made more perfect. Here we shall have to follow a path that is the direct opposite of that described in the last lecture and followed in my book on mechanics. The abstract concepts of material point, mass, force and so on which were there our starting-point will of course not be entirely avoidable here either, for they simply are the basic pillars on which mechanics rests; but we shall now introduce them as late as possible and while previously they were postulated, now they will be tied as far as possible to experience, from which we shall endeavour to deduce our results. For this reason we must not now begin with those laws that previously seemed the simplest, such as the law of inertia, whose usual formulation is that a material point free from all external influence moves uniformly in a straight line. Apart from the difficulty inherent in the concept of a material point, we cannot remove any body so far from all others that it is free from all influence, and if it were possible then we could no longer observe its motion let alone that it moves uniformly in a straight line. However, in order to verify the law of inertia for actual bodies the forces acting on which are in equilibrium, one would have to give a prior account of the whole theory of equilibrium. Thus in the ordinary presentation abstractions and facts are usually intermingled, which it is precisely our main task to avoid in what follows, since we propose to start strictly from pure facts of experience.

The first inconvenience that meets us here is this: previously when setting up the basic principles we were dealing with purely mental constructs which we can mentally transform as we like and of which we can require that they always conform exactly to our demands; whereas now we wish to start from phenomena directly observed, which must alway be very composite and complex. If from them we wish to obtain fundamental laws, we always must generalize and idealize the phenomena, so that we really are dealing no longer with what is quite exactly factual but with processes that are realized in nature only more or less approximately. Therefore we cannot quite avoid mixing ideas and facts, although we do our best to reduce this to the least possible measure and try not to do it furtively but to remain clearly conscious of our actions where we are forced into them.

Phenomena that are given to us are of extremely various kinds. The simplest consist in a body's change of position, while its shape and other properties seem to remain the same. Already this simple phenomenon is in some ways idealized. Only in the rarest cases is there absolutely no change in a body's shape; indeed, all bodies, even the most unchangeable, can break under very strong forces, and heat or chemical action may provoke a total change of their properties. Still, there are very many bodies whose shape does not noticeably alter during the most varied motions over long periods of time. We call them solids and so form the ideal of an absolutely immutable body which we call rigid. Other bodies, namely fluids, alter their shape in the most varied ways while moving, either with constant volume (approximately, of course) as in liquids or under constant and very noticeable changes of volume, as in gases. This last phenomenon can be reduced to the former by assuming that fluids consist of very many small particles that move independently of each other and so provoke the change of shape. If the average distance of two neighbouring particles changes, a change of volume occurs as well. There is now the problem whether we are to consider the number of these particles as mathematically infinite or merely very large but finite. Many facts of experience point to our having to assume the latter, which happens to be philosophically more satisfying as well. However, since the question has not so far been settled by a decisive experiment, we shall leave the question entirely in suspension, in accordance with the principles we now propose to follow.

All changes of position are called motions. The theory of the phenomena of motion is mechanics, which is subdivided into mechanics of rigid bodies (solids), hydromechanics (liquids) and aeromechanics (gases). According to its definition, mechanics further comprises the conditions under which a body does not move at all.

There are many other kinds of phenomena, like sound, heat, light, electricity, magnetic phenomena, the total alteration of the properties of bodies in chemical processes, smell, taste and so on. These last are proably only special cases of evaporation or chemical phenomena and therefore of little significance to physics, which leaves action on nerves, and transmission by them into consciousness, to physiology and psychology. Nevertheless, they must be mentioned here.

It is established beyond doubt that at the basis of phenomena of sound there are motions of bodies. Accordingly scientist similarly have tried to explain light, electricity and magnetism as well as chemical phenomena by motions of certain hypothetical media or smallest parts, so that until recently most physicists were no doubt convinced that therein consisted the essential task of physics. Only a few decades ago irrefutable proof was given that the theory of electric and magnetic fluids, especially popular in Germany before that time, was irreconcileable with the facts. Now people became more cautious; they continued at first to try to explain electric and magnetic phenomena by the mechanical action of a medium, but since this did not lead to definite and unambiguous results, some physicists have lately inclined to the view that it was premature to regard all phenomena as explicable in terms of motions. Using our terminology, it might perhaps be impossible to form an adequate picture of phenomena by means of the pictures of points changing position alone, a variety of other pictures of various qualitative kinds being also required, such as dielectric and magnetic polarisation, chemical states or whatever. This would seriously impair the unity of science, since it would certainly be impossible to avoid the old and simple pictures, on top of which a lot of extraneous ones would have to be introduced. Moreover this would call in question the significance of mechanics as the basis of natural science as a whole, all other scientific theories resting on mechanics. Yet mechanics as the theory of the simplest phenomena without which no others could be conceivable, would still have to come before all other physical theories. If therefore on the one hand it has indeed so far been impossible to prove that all natural phenomena can be mechanically explained, on the other it is equally certain that nobody has shown certain phenomena to be not thus explicable: at most we can say that for certain phenomena an attempt at mechanical explanation is as yet premature. The general question as such can be resolved only after centuries, or at least be given a new setting and clarified. We shall therefore not tarry to discuss the 'pros' and 'cons', but return to the motion of a solid body which will from the outset be idealized and considered absolutely rigid. We do not conceive it as a material point, but as a body given in experience, apparently continuous and extended. We must again start with an immediate abstraction: we cannot grasp the body's motion as a single whole, since (at least apparently to us) the body consists of infinitely many parts. It is only the motion of its individual points that we can clearly follow in sight and thought. Let us therefore denote very small positions on it with delicate markers that are of course rigidly connected with the body, let us say small coloured spots or specks of flour or intersections of two thin lines and so on. If we sink very fine bore holes into the body we can actually mark internal points as well, and we can do so in any case in thought without holes if we conceive of a geometrically similar body either hollow or transparent or otherwise accessible at the point in question. It is of course a further idealization to regard these marked positions as mathematical points, but we stay much closer to reality if we describe the motion of extended bodies by such points, trying in the first place to obtain simple laws for the mechanics of extended bodies, than if we start directly with the laws of motion for individual material points. We can now more accurately describe what is meant by saying that the shape of a body does not change during motion. By applying a measuring rod or two compass points subsequently transferred to a rod, we can measure the distance between any two points of the body, that is between any two of the markers; if this distance remains the same for all point pairs at all times, then we say that the body's shape is unchangeable. However, we have of course no objective guarantee that the measuring rod or the pair of compass points remain unchanged, all we can say is that experience has given us the correct result in all cases when they were applied to bodies whose shape seemed unaltered to our eyes. If with time all solid bodies were to change their dimensions similarly, we could of course not notice it. Nor do we intend to explain why there are solid bodies and why we can measure the distances between markers rigidly connected with them. We take these as given facts of experience. What we wish to represent by means of our mental pictures is merely the change in distance between markers of different bodies or of the same body if it is not rigid.

A precondition of any scientific insight is the principle that natural processes are unambiguously determined; or, in the case of mechanics, that motions are. This principle asserts that the motions of bodies are not haphazard, occurring now this way now that, but rather that they are unambiguously determined by the circumstances, in which a body finds itself. If every body moved at random, if under given circumstances now this now that motion occurred by chance, then we could at best look with curiosity at how things run, but not investigate their course. Here again there is a lack of definition, for the circumstances under which any body moves comprise strictly speaking the whole universe, which is never in the same state twice. We must therefore reduce our conditions and require that the same motion always occurs if the immediate surroundings are in the same state. Here the inductive method is again much less advantageous than the deductive. Since in deduction we begin with stating the various laws of action irrespective of any experience, we are at first entirely free to determine arbitrarily which circumstances affect the motion of a body and which do not. With induction on the other hand we must determine from experience the concept of a body's immediate surroundings whose state has influence on its motion. On the theory of contiguous action it is only immediately adjacent volume elements that influence the motion of any given such. Thus, on this theory the Earth does not directly attract gravitating bodies but acts merely on the volume elements of a medium that transmits the effect to the body. However, if we are to remain faithful to the principles of our present method of representation, we must not make contiguous action the basis of the whole edifice of mechanics: for that we must use only such laws as contain no arbitrary features, being on the contrary forced on us unambiguously and necessarily by experience. The theory of contiguous action, on the other hand, however a priori likely it may seem to some, still goes completely beyond the facts and to date remains well beyond what can be elaborated in detail. We should fall into the same error that we have laid to the charge of Hertz's mode of representation: either we should have to invent quite arbitrary special hypotheses for the way in which contiguous action operates, or make do with vague general notions about it all.

We must therefore include the whole Earth as part of the surroundings of a gravitating body, but leave the Moon and stars out of account, since they have no noticeable influence. It is thus once again a pure assumption, to be subsequently justified by experience, that we can always draw the boundaries of immediate surroundings in such a way as to include all essentials, and thus actually arrive at a formulation of laws of motion.

What will be our attitude to absolute space and time, given our present mode of representation? We cannot put compass points on parts of absolute space, but only on material bodies. Hence we can determine only the motion of material bodies relatively to each other. We must not at this stage mix the real bodies we are now alone considering with the mental picture of a fictitious co-ordinate system, as we did with the deductive method. Rather, following the spirit of our method, we must link our consideration as closely as possible with the historical development of mechanics. Galileo found the simple laws of motion by studying motions relative to the Earth. Following his example, we shall include in our considerations not only the body whose motion we wish to describe but also a system of other bodies satisfying the condition that all their points retain their relative distances; that is, all of them are at relative rest. This system we call the reference frame. If therefore we are studying the motion of a solid body with regard to a reference frame and if $A, B, C...$ are marked points of the former and $E, F, G, ...$ of the latter, then neither the distances $AB, AC, ...$ nor $EF, EG, ...$ change, so that our problem consists merely in establishing the laws of change of the distances $AE, AF, BF, ...$, which of course again requires much idealization. We shall hardly find a system of bodies as reference frame such that their relative position remains absolutely the same at all times. It is enough if this is approximately so for sufficiently long periods.

Moreover, we cannot know whether we shall obtain the same laws if we choose one reference frame rather than another. We shall thus have to choose one that yields simple laws of motion. It turns out in fact that the laws obtained by choosing the fixed stars as reference frame cannot without considerable corrections be applied to motion relative to the Earth, and we must regard as an extraordinarily favourable accident that the effect of the Earth's rotation on the various types of motion observed on its surface are so very small. Otherwise it would have been much more difficult to derive the fundamental laws of mechanics. It is owing to this circumstance that for motions on Earth we can choose the Earth itself as reference frame. This yields simple laws that do not indeed describe actual motions with absolute accuracy, but the deviations are so slight as almost to defy observation. Of course, we could not know this a priori, but it is no logical mistake if we begin by studying the laws of motions relative to the Earth. If we find simple laws, it is again no logical mistake to try applying them to the motion of the planets relative to the fixed stars. It is through this extension that we first find on the one hand that for the former case too the laws must hold approximately, and on the other that nevertheless small corrections are then necessary. These are so small as not to interfere with our previous discovery of the laws from terrestrial motions, but now that we have recognized their order of magnitude they are nevertheless observable with more delicate devices. The fact that actual motions then show precisely the peculiarities caused by these corrections justifies our method a posteriori in most brilliant fashion. Thereby we once more eliminate the pedagogic difficulty caused by the relativity of all motion. The question as to which reference frame must be chosen for the motions of the fixed stars is of course not solved by this, but it is by no means necessary to treat this question ahead of establishing all laws of mechanics.

So far we have made no special assumptions as to the shape and arrangement of the bodies of the chosen reference frame. There is however no difficulty about conceiving them linked with three mutually orthogonal straight lines, which can be chosen as co-ordinate axes. The position of each marked point of the body in question is then always determined by its rectangular co-ordinates in that co-ordinate system. If these do not change with time, then the body is at rest in that frame. If they do change, the body moves, in which case in order to give a description we must still fix the unit of time. Just as we distinguish larger spatial extensions from smaller ones just by means of sight and touch but can gain a numerical expression for spatial size only by comparison with a rationally constructed scale, so likewise we can distinguish longer from shorter periods by our sense of time but must obtain a precise quantitative measure of time using the means indicated earlier in my first lecture. Above all we must obtain a series of processes that give us a perfect or rather the best possible guarantee that they occur in equal times. For example we might drop identical pendulums from rest by identical distances. When the first reaches its rest position we release the second and so on. Whether we have in fact sufficiently avoided mutual interference can of course be shown only by comparison with various similar trials. We naturally soon perceive that a single pendulum on its own performs the various consecutive oscillations almost under the same conditions and we can use them for measuring time. Again of course absolute isochrony of oscillation is an ideal, since temperature, pressure, sun and moon have some influence on it; but how all these interfering circumstances are largely avoided in carefully built chronometers, how a driving impulse keeps the oscillations going for long spans of time, that when a certain chronometer finally becomes unusable another as far as possible like it can be substituted, all this no longer concerns our present general considerations.

We choose a certain moment in time as zero, for example the time of an arbitrarily chosen transit through the rest position, and take the next such transit as time 1, and subsequent ones as 2, 3 and so on. Subdivisions can be determined by a tuning fork that oscillates faster or by means of motions that have proved to be sufficiently uniform under all circumstances over large intervals so that we have good reasons to surmise the same for smaller intervals. In this way we obtain the times $\frac{1}{2}$, $\frac{1}{3}$ and so on and we can never fix a limit of subdivision. Negative numbers denote oscillations previous to that chosen as corresponding to zero. In this way we can represent all times by positive, negative, whole, fractional or irrational numbers, just as we represent lengths by the number that indicates how often they contain the unit of length. The difference of two numbers representing two given times is called the time interval between them or the time difference or the time elapsed meanwhile. Our ordinary unit of time is derived from the Earth's period of rotation, whose uniformity for the derivation of the principles of mechanics is better tested by simpler processes, since without a knowledge of mechanical laws it is not immediately obvious that the velocity of rotation remains the same at all points of the Earth's orbit.

We now return to our rigid body, relating it to a co-ordinate system Ox, Oy, Oz rigidly connected with the chosen reference frame. Consider some marked point on it which at a definite time t is at A with rectangular co-ordinates x, y, z. Join A to the origin of co-ordinates O by the line OA, which is called the position vector of A, its projections on the three co-ordinate axes being the three co-ordinates x, y, z. If now the body undergoes a certain given motion, we must first represent every moment of time during the motion by a number, let us say by means of comparison with the simultaneous motion of our chronoscope. To every time there will correspond a definite position of the body and therefore of the point A and thus

definite values of the co-ordinates x, y, z which we also conceive as represented purely by numbers, that is whole or fractional multiples of the
unit of length. To every numerical value t of time thus belongs an unambiguously determined value of the co-ordinate x, which is an unambiguous function of t, and likewise for y and z. These we write $x = \phi(t)$,
$y = \chi(t)$, $z = \psi(t)$, calling t the argument or independent variable and
x, y, z the dependent variables. We can at first take it as a fairly certain
fact of experience that a body never suddenly disappears from its position
reappearing at the next moment in another position at a finite distance
from the first, and likewise for each part of a body, so that ϕ, χ, ψ are
continuous functions of time, that is their increments become vanishingly
small for vanishingly small increments of time. The curve formed by the
various positions of A at different times we call the path of this point, and
that part of it which corresponds to all positions traversed in a given time
the path travelled in that time.

What is less certain is whether the continuous functions ϕ, χ, ψ are
differentiable as well. In traditional mechanics it used to be put thus: let
a point of a body traverse a very small path δs during a very small time δt,
and let it be a priori clear that during this small time span the circumstances
in which the body finds itself can have changed only very slightly, so that
during the next time span δt it will again traverse a path very nearly equal
to and in the same direction as δs; thus for small times both path and
co-ordinate increments must be proportional to the time elapsed. In those
days it was in any case generally believed that every function that is everywhere continuous must have a differential coefficient. As is well-known,
Weierstrass has shown this to be an error: let y denote Weierstrass's
series, then the increment of y corresponding to any increment of x tends
everywhere to zero if the latter does and yet their ratio never tends to a
determinable limit. With the deductive method this does not cause the
slightest difficulty, for we can form our picture as we wish and include differentiability from the outset, justifying this in terms of subsequent agreement with experience. However, our present intention is to start from
experience. This does indeed tell us that in many cases during small but
still observable time spans the path of a point of a body is the more precisely proportional to the time elapsed the shorter the latter, from which
we may well infer that ϕ, χ, ψ are differentiable. But we are also aware of
examples of very rapid oscillations and cannot prove exactly whether in

certain cases there might not be motions that are better represented by a series resembling Weierstrass function than by a differentiable one. Still, these matters are of slight importance, so that we are going to base our further considerations on the assumption that the co-ordinates are differentiable with regard to time. With this presupposition the functions ϕ, χ, ψ have derivatives with regard to time, and these we call the components of the velocity of the point A of the body. The velocity itself can be constructed as follows: let the marked point of the body be at A at time t, at B at time $t + \delta t$, so that OA and OB are the two corresponding position vectors. Then the line AB is called the difference between the two vectors. We now construct a vector of direction AB and length equal to the ratio of AB to δt, and then seek the limit which this vector approaches in size and direction when δt tends to zero. The length thus determined is the size of the velocity and the limiting direction its direction. One further comment: to be able to divide the path by the time elapsed both must be expressed by pure numbers and we saw how this is done. If we choose a unit of length a times greater, the number expressing a given length becomes a times smaller. There may be other quantities that have this property, that they appear expressed by numbers a times smaller if the unit of length is chosen a times greater. Of all such quantities we then say that they have the dimension of a length. Each length (path, co-ordinates and so on) thus obviously has the dimension of a length. The number that expresses the time t is of course independent of the unit of length chosen, but it becomes a times smaller if we choose the unit of time a times greater and we say of any quantity that is expressed by a number of this kind that it has the dimension of time. Velocity is measured by the ratio of two numbers the numerator having the dimension of length and the denominator of time. It therefore depends on the choice of units both for length and time, becoming a times smaller if the former is chosen a times greater but a times greater if latter becomes a times greater. We therefore say that velocity has the dimension of length divided by time, although this is completely divested of any mystery or metaphysical significance. One often simply speaks of the ratio of length to time instead of speaking of the ratio of numbers expressing them. Thus we have extended the concept of division so that the ratio of length to time must be given quite a new definition, just as the concepts of negative or fractional powers must be newly defined, as a fraction or root respectively. The advantage of this new definition is that we can largely take over the rules of calculation established for the previous definition. However, we must not infer a priori that this holds for all such rules, on the contrary it must be separately proved for each rule. It is equally quite a new definition if by the second or third power of a centimetre we understand the geometrical figures of a square or cube, respectively, of side 1 cm, and we must justify how far this new definition is appropriate. At this stage there remains not the least difficulty in fixing the concept of acceleration and its components in the three co-ordinate directions. Let OC be the vector representing velocity in size and direction at time t, and OD at $t + \delta t$, so that CD is the difference between them, which will be very small for small δt. However it will remain a finite line if we multiply it by the ratio of $1/\delta t$ leaving the direction the same. The limit of this augmented vector for vanishing δt is called the acceleration vector, its length representing the size and its direction the direction of the acceleration. Its components in the three co-ordinate directions are called the components of acceleration. We readily see in the usual way that they are the second derivatives of our functions ϕ, χ, ψ. We must therefore presuppose that these functions have second derivatives as well as first. It is further easy to see that the number expressing the size of the acceleration again depends on the chosen units of length and time, becoming a times smaller if the former is made a times greater, but a^2 times greater if the latter is made a times greater. We shall therefore say that the dimensions of acceleration are length divided by the square of time. We can again define acceleration as such, as the ratio of a velocity to a time or a length to the square of a time, though we must use these definitions with some care since they represent extensions of the concept of algebraic division for which we must first test afresh whether the various rules established in algebra are applicable.

Having developed these concepts as far as possible in connection with experience, we must proceed to establishing the laws that govern the motions of bodies. Here again we shall of course not begin with the laws for a material point, since that is an abstraction, although we shall likewise refrain from entertaining the illusion that we can do without all abstraction. In my view we cannot pronounce a single proposition that would really be merely a pure fact of experience. The simplest terms like yellow, sweet, sour and so on which seem to indicate mere sensations already express concepts obtained by abstraction from a great many facts of experience. When Goethe says that only half of our experience is experience, what he intends to convey by this seemingly paradoxical dictum is surely that in every conceptual grasp of experience or verbal representation of it we must already go beyond experience. The frequently uttered requirement that natural science must never go beyond experience should therefore in my view be reformulated thus: never go too far beyond experience and introduce only such abstractions as can soon be tested by experience. We shall not put the law of inertia at the beginning either. It may be the theoretically the simplest law of mechanics, but physically it is by no means so since it presupposes a whole series of abstractions, as pointed out earlier. Rather, what appear to us as the two simplest physical cases are that of relative rest and free fall of heavy bodies. As we saw, we can never isolate a body from all external influence. If now there are such influences and each of them on its own would produce a motion but their joint effect produces rest in the reference frame, then we say that the several causes of relative motion cancel each other. Or I might use the common expression that the forces are in equilibrium, but I will deliberately avoid the usual expressions since we naturally link them quite unwittingly with a host of notions which then insinuate themselves unwanted and unchecked into our arguments thus provoking a semblance of proof when we have merely introduced without rational backing certain notions that correspond to our old mental habits and associations. Moreover, I will avoid the term force until I can speak of mass at the same time. Finally, we shall here consider only relative motion. However, a body may be at rest relatively to its surroundings without the forces acting on it being in equilibrium, for example a body which is at rest in a lift that moves under acceleration.

Consider now a specific case where the causes of relative motion cancel. Let a heavy body be suspended by a thin thread. We might think that here there are no causes of motion present. However we find that motion supervenes as soon as we remove the thread, so that there must have been at least two causes of motion that cancelled each other.

If we analyse the motion that occurs on removal of the thread we find that given certain general conditions it will very nearly always proceed in the same manner. The general conditions are these: the body's surface must not be too big compared with its weight, there must be no violent movement of air round the body, the thread must be cut without shock or gently burnt through or destroyed in some other way. The same motion occurs if we start by holding the body by hand or with pliers or some other device and then suddenly release it without shock. The characteristic feature of all these initial conditions is that all points of the body have very small velocities during the first moments of the motion. We can therefore approximately assume that at the first moment of the motion all points of the body had zero velocity. If these conditions are fulfilled, experience shows that the body moves almost exactly according to the same laws wherever it might have been released near the Earth's surface. For the time being the motion will of course be determined relatively to the Earth. If we further confine ourselves to not too large a portion of the Earth's surface, the direction of motion too is everywhere the same, namely that of the thread initially supporting the body. Experience now shows that this motion is governed by the following laws. Firstly, the body moves parallel to itself, that is in a given time all its points move through the same distance in the same direction; since therefore each point has the same path we can denote this as the path of the body. Secondly, all these paths are straight lines. Thirdly, the velocity grows steadily, but the acceleration is the same everywhere and everywhen, indeed for all bodies. That these laws are realized in nature only more or less approximately has already been discussed.

We can now repeat the same experiment except that at the time of release we give the body an impulse or otherwise ensure that it will have an initial velocity other than zero. Since we have not yet met the theorems concerning the centre of gravity and the rotation of bodies, we must confine ourselves to the cases where the body once again moves parallel to itself. This will not always happen nor can we at this stage give the conditions that it should occur, but in many cases it will happen and for the present we shall consider these alone. In all these cases the points of the body again traverse equal paths which we can therefore call the path of the body. The whole motion can again be described as follows: the acceleration is always vertically downwards and everywhere the same for all bodies. Since we have now seen that the motion will always proceed in the same manner, wherever we have allowed it to start in a room or its near surroundings, we must conclude that the cause of motion, which we call force, is the same and constant in all those places. Besides the acceleration is also constantly the same, so that we further conclude that at least in this special case acceleration is the decisive feature of force and since the former is everywhere directed vertically downwards we say that the body is acted on by a constant vertically downward force, namely gravity.

ON THE PRINCIPLES OF MECHANICS*

PREFACE

In response to repeated requests I here publish two inaugural lectures held at the beginning of my teaching assignment at the University of Leipzig and then on resuming my old professorship in Vienna, although I am convinced that those who have not previously heard these lectures will be greatly disappointed.

They will expect a new edition of the philosophical criticisms that I once presented to a meeting of natural scientists in Munich. They fail to consider that on that occasion my audience was quite a different one: there most of them were men who if not initiated into all the details of theoretical physics were nevertheless surfeited with science so that they might well have desired the digestive tablets of critical philosophy.

But what could these mean to audiences like those I expected for the following two talks? These consisted of young men who were eager for science and wanted first to absorb scientific theories rather than give them out again; they desired not so much a savoury to promote digestion as an hors d'oeuvre to whet the appetite.

At least that was my view, whether correct or not. In any case the two following pieces are to be judged with lenience: do not expect any deep thoughts, but merely harmless small talk.

I. LEIPZIG, NOVEMBER, 1900

When we propose to introduce new guests into the home that we have long inhabited, we customarily deck out the entrance door with festive ornament. I have been called to this ancient and venerable university in order to introduce you into the vast and imposing edifice of theoretical physics. The entrance gate through which we step into this edifice is

* *Populäre Schriften*, Essay 17.

analytical mechanics. No wonder then that I wish to show you that science in its most beautiful finery with which it has been adorned not by me but by the choicest spirits over the centuries.

As a true theoretician, I wish to consider the central core rather than all the external detail. The definition of analytical mechanics is very simple: it is the theory of the laws governing the motion of bodies. A knowledge of these laws is required for the treatment of many machines and similar devices whose simplest forms were known already in the dim and distant past, to Egyptians and Babylonians. It is therefore not to be wondered at that the first beginnings of exploration of mechanical laws go a very long way back. Although it was almost always a matter of setting bodies into motion researches, except for a few unsuccessful attempts, were confined until Galileo's time to the conditions of equilibrium, which in the cases examined in those days coincided with the conditions under which bodies did not move at all. It is odd that consideration of this case, which does indeed fall under our definition of mechanics but only as a special instance, or almost an exception, should suffice for dealing with the machines then in use; but since this amounts to ignoring precisely that which should be described, namely actual motion, no science of mechanics in the proper sense had yet been attained. Such a science does not begin until Galileo, who by experiments that were in equal measure ingenious and fundamental ascertained once and for all the basic laws for the simplest cases of motion.

One might have expected that these laws could next be applied to more complicated terrestrial phenomena, for instance the growth of a blade of grass, and thereby be extended, but this was by no means the case. These and similar terrestrial processes that seem unremarkable to the naive observer are as yet complete riddles to us. Progress was much rather initiated by Newton's at once applying the basic laws discovered by Galileo to the motion of what is most remote from us, namely celestial bodies; for it was precisely on this path that Newton found those extensions and completions of Galileo's laws which in turn could be applied to more complicated terrestrial motions, so that he succeeded in working out a theory of the motion of bodies so perfect that to this day it has become the foundation not only of mechanics but of the whole of theoretical physics.

On this Newtonian basis the most outstanding analysts of all nations built further, amongst them Lagrange, Laplace, Euler, Hamilton and so analytical mechanics gave rise to a creation which is rightly admired as a paradigm for any mathematical physical theory.

The first result achieved was the formulation of the laws of motion for rigid bodies in equations, so that all such problems can be reduced to a pure matter of calculation.

However, mechanical ideas were also developed as to the internal structure of solids and fluids, leading to equations that express the laws of elasticity for the former (deformation, rigidity) and of motion for the latter. Now as soon as a field of phenomena has been formulated in equations the physicist regards his task as done, their solution he delegates to the mathematician. How far we are from actually solving all these equations, that is being able in all cases to obtain from them a genuinely clear picture of the processes in question, a simple glance at a foaming brook or at the water waves churned up by a large steamer will tell. How impotent analysis really is to read from the hydrodynamic equations the details of all these phenomena! And yet mechanics in all these fields provides formulae that are of inestimable practical value for the construction of buildings, iron bridges and towers, canals, pumping stations and so on, not to mention the countless machines that daily not only replace but even excel manual work in an amazing fashion.

The ability to think in mechanical terms is of supreme utility in all spheres of practical life and has a formative and educating influence on the whole of intellectual life. Just as a good teacher endowed with proper psychological knowledge treats every one of his fellow-men in just the way that the individuality of each requires, so the person who thinks in mechanical terms meets every mechanism, from the simplest to the most complex, with love and respect, and it is worth it because the mechanism fulfils its master's wishes, whereas the mechanically ignorant does not even notice in which sense a screw must be turned, thus inseparably joining what he wanted to take apart.

If a nation has achieved great results in comparison with its neighbours, it tends to attain a certain hegemony over the latter and not infrequently proceeds to render them subject and subservient. The same happens with scientific disciplines. First it was acoustics that surrendered quite naturally and without resistance. The phenomena in question are intimately linked with phenomena of motion that are of course so fast that we cannot follow them directly by eye, but without concealing their purely kinetic character even on merely superficial observation. Indeed by artificial means both the motion of sound-exciting elements as well as of sound-waves transmitted in air can be made visible and recognizable. Thus, acoustics was immediately claimed by mechanics as its domain. The same happened with optics when it was recognized that light like sound is a phenomenon of waves and oscillations. The construction of an oscillating medium was of course left entirely to the fancy and encountered not inconsiderable difficulties.

The campaign into the territory of heat theory was opened by mechanics with the idea that heat is a motion of the smallest particles of a body; remaining invisible because they are so small, but making itself felt by provoking the sensation of heat when it is imparted to the molecules of our body, or of cold when it is withdrawn. This campaign was victorious because the hypothesis described furnishes a very clear picture of the agency called heat, much more completely so than the earlier view of it as behaving like a substance.

Electricity and magnetism were subsumed under mechanical laws by means of the hypothesis of electric and magnetic fluids whose particles were supposed to be acting on each other according to a law that was merely a variant of Newton's law for the mutual action of celestial bodies and thus rests firmly within purely mechanical territory. Finally much success was achieved in trying to reduce chemical phenomena and the formation of crystals to a mechanics of attracting and repelling forces and of the mutual motion of heterogeneous atoms; indeed, chemical phenomena are closely connected with thermal ones as well as with electric ones. About an opposing tendency more recently directed against these theoretical endeavours, more below.

Even the most superficial observation shows that the laws of mechanics are not restricted to inanimate nature. The eye is an optical dark chamber down to the minutest details, the heart is a pump, the muscles a complicated system of levers intelligible only from the point of view of pure mechanics capable of solving seemingly most intricate problems by the simplest means. For example, all conceivable movements of the eye are achieved by six muscular cables acting like threads pulling a sphere that can move about its centre of gravity; of course, the full expression of opening the eyes or lowering the glance of which novelists tell, is caused also by the external decoration – the play of the eyelids and facial muscles and other things.

The applicability of mechanics reaches much further into the mental sphere than might appear at a superficial glance. Who, for example, has not made observations that provide evidence for the mechanical nature of memory? At one time it often happened that in order to remember a certain Greek word I had to recite a whole series of memorised Homeric verses, when the word immediately came to mind at the appropriate spot. During a period when for weeks on end I worked on Hertz's mechanics I once meant to start a letter to my wife with the words 'Dear Heart' and before I knew I had spelt the word with 'tz'.

Everybody knows how often the innate alarm clock built into our memory leaves us in the lurch if it is not supported by special mechanisms (a knot in the handkerchief, hanging the umbrella over the winter coat). On the day of my move to Leipzig I went to the window in order to read the thermometer as usual, though I had myself unscrewed it the day before. I was driven to exclaim: "I possess no other mechanism that functions as badly as my memory, except perhaps my judgment".

Thus we may see in our bodies an ingenious mechanism and its diseases too can be explained by purely mechanical causes. This insight has been extremely useful already, by showing the way and goal for the surgeon's mechanical interventions, by uncovering the true mechanism of infectious diseases and so avoiding them by keeping away the bacteria that cause them or curing the disease by killing the bacteria. However, in most cases we are still powerless when confronted with the violence of nature, but mechanics still helps us to understand it and thus to bear it too.

We must mention also that most splendid mechanical theory in the field of biology, namely the doctrine of Darwin. This undertakes to explain the whole multiplicity of plants and the animal kingdom from the purely mechanical principle of heredity, which like all mechanical principles of genesis remains of course obscure.

The explanation of the exquisite beauty of flowers, the great wealth of forms in the world of insects, the appropriate construction of organs in human and animal bodies, all this thereby becomes a domain of mechanics. We understand why it was useful and important to our species that certain sense impressions were flattering and therefore sought after, while others were repellent; we see how advantageous it was to construct the most accurate pictures possible of our surroundings and strictly to keep apart that which coincided with experience as true from what did not as false. Thus we can explain the genesis of the concept of beauty, just as that of the concept of truth, in terms of mechanics.

Moreover, we understand why only those individuals could continue to exist which abhorred certain very noxious influences with all the nervous energy at their command and sought to keep them in the background, while with equal vigour aiming at other influences that were important for the preservation of themselves or their kind. In this way we grasp how the intensity and power of our whole affective life developed, pleasure and pain, hate and love, happiness and despair. Just as we cannot rid ourselves of bodily disease, so likewise with the whole gamut of passions, but again we learn how to understand and bear them.

In the first instance it will without question be important for any individual that his endeavours are turned towards his own preservation, so that egoism appears not as a defect but as a necessity. Yet for the preservation of the species it is of the greatest utility if the different individuals help each other and that in collaboration each subordinates himself to the whole. For example, we understand the need for wilfulness and stubbornness in children, but also for solidarity and sociability during play; we understand the human features of self-interest and sympathy, shame and desire, love of freedom and sense of submission, virtue and vice, fear and contempt of death. It is of great advantage to united action in peace and war if young men are inspired to great and noble things, friendship and love, freedom and patriotism, but how easily this drive degenerates into empty phrases and inactive enthusiasm. A receptiveness for exaltation and inspiration thus had to form itself in our species but so did sobriety and egoism, as a kind of counterweight. In this way, we understand the mechanical reasons why one youngster is aglow with the poetry of Schiller while many condemn the poems of Heine, which nevertheless have a powerful and irresistible influence on others. The rising water in a fountain must indeed have enough kinetic energy to propel it into infinite space, but with equal mechanical necessity we have the reaction of gravity and the pressure of countless air particles which ensure that it returns to mother Earth. To put it picturesquely, one might go so far as to assert that not only the most abject vice but also the highest virtue is in a sense an aberration, grounded in the fact that our innate drives aim beyond the target. For excessive idealism clouds practical sense and is thus just as harmful as a banausic mentality, the opposite extreme. Such paradoxes are more obvious than one might think and, like the distorted pictures in a cylindrical or conical mirror, they arise whenever one considers things from a one-sided position. Similarly it has been said that genius is a mental illness.

Indeed man cannot even claim the ideal for his own species alone. By whipping a dog when unfaithful but feeding it in the reverse case, man has bred faithfulness in dogs just as he has bred cows for high yields of milk or geese for large livers. The more attached dog was always favoured by man in its struggle for existence so that in the canine species attachment and fidelity grew to an ever greater degree. If now as often happens, a dog who has lost his master no longer eats and slowly dies of grief, is not that a kind of idealism that we hardly even find amongst men and certainly not amongst our modern servants? For this reason many philosophers have been tempted to rank dogs morally higher than man, just as one might be tempted to place the automatic art of nestbuilding of birds above the laboriously acquired art of the architect, which is subject to mistakes.

In nature and art the all-powerful science of mechanics is thus ruler, and likewise in politics and social life. Because of the tremendous instinct for independence of which we saw that it must necessarily develop even in the child, the individual is not willingly dominated by others but likes a republican form of government in social groupings, cities, communities and states. Against this, however, other mechanical difficulties stand opposed. Anyone who has assisted at public debates knows how unsuitable and clumsy a public meeting is as an organism for quick and consistent action and how often it makes faulty decisions because each individual bears only a slight measure of responsibility. This is the more likely to happen because of a circumstance that Schiller describes in these words: "Judgment has always belonged to only a few". These reasons illustrate, on the other side, the advantages of the rule of few or one alone. Thus the collaboration of the most varied personalities in popular assemblies, as well as an individual's masterly direction of the mass and its recalcitrant willfulness, both these in fact rest on the mechanics of psychology. Bismarck could see through his political opponents' thinking as clearly as a mechanical engineer sees through the gears of his machine, so that he knew exactly how to move them to act as he wished just as the machinist knows what lever to push. The enthusiastic love of freedom of men like Cato, Brutus, and Verrina arises from feelings that had grown in their souls by purely mechanical causes and we again can explain mechanically that we live contentedly in a well-ordered monarchical state and yet like to see our sons reading Plutarch and Schiller and draw inspiration from the words and deeds of enthusiastic republicans. This too we cannot alter, but we learn to understand and bear it. The god by whose grace kings rule is the fundamental law of mechanics.

It is well-known that Darwin's theory explains by no means merely the appropriate character of human and animal bodily organs, but also gives an account why often inappropriate and rudimentary organs or even errors of organisation could and must occur.

It is no different in the field of our drives and passions. By adaptation and heredity only the basic drives could develop which are on the whole necessary for the preservation of individual and kind. Here it is unavoidable that in some few cases the basic drives act wrongly, and become useless or even harmful. Our innate drives often overshoot the mark as it were. The force with which for the sake of certain results they have linked themselves with our minds is so enormous that we cannot easily rid ourselves of it if these results are not achieved and the now habitual drive is superfluous or harmful. For example, for the new-born child the sucking instinct exceeds all others in importance; no wonder then if it exceeds them in intensity as well and later becomes embarrassing when the by now rational child often cannot shed it for an incredibly long period. Grownups smile at this and yet the inappropriate and incorrect persistence of the instinct that serves the preservation of the species not infrequently takes on much crazier forms still.

Similar phenomena exist in the purely mental sphere. For example we have associated our feelings so much with certain ideas and impressions, that a cleverly constructed fictional tale or a stage play moves us much more than a short and accurate report of an actual mishap to people that are distant from us.

In philosophical thought similar effects occur. We are accustomed to judge the value or otherwise of various things according to their help or hindrance to our own existence. This becomes so much a habit that in the end we imagine we can judge as to the value or otherwise of life itself, and indeed whole books are written about this mistaken topic.

In my view the laws of thought arose from the fact that the linking of internal ideas that we form of objects became more and more adapted to the actual links between objects. All rules that led to contradiction with experience were rejected and those that always led to what was correct were so fiercely retained and the retention so consistently passed on by heredity that in the end we saw in such rules axioms or innate mental necessities. However, even here in logic one cannot exclude overshooting the target. Indeed, precisely because this field is so abstract and seemingly perspicuous it makes fools of us most in such cases. This in my view is the origin of those contradictions that are called antinomies in Kant and riddles of the universe in more recent times. Let me mention some examples. We must constantly dismantle concepts into simpler elements and explain phenomena by means of laws we know already. This highly useful and necessary activity becomes so much a habit as to produce the compelling appearance that the simplest concepts themselves must be dismantled into their elements and the elementary laws reduced to even simpler ones.

Questions like what is the definition of the number concept, the cause of the law of causality, the nature of matter, force, energy and so on, always irresistibly recur, even to the person who is philosophically trained. He is convinced that these concepts are taken straight from experience and not explicable further, so that here the now irresistible mental habit of asking for the cause and definition overshoots the mark, but still he cannot overcome a certain residual dissatisfaction that such important concepts as number or causality defy all attempts at definition. It is as when an optical illusion fails to vanish even after one has clarified its mechanical cause.

It is a step further still if we find it inexplicable and mysterious that we or anything at all should exist and cannot quite rid ourselves of this notion even after recognizing that the concept of mystery is here as little applicable as the concept of value or otherwise in judging life as a whole.

Another example that belongs here is that old aberration we now call solipsism. Just as it is mechanically explicable that a surge of blood in the ear can produce the sensation of a note or that we perceive after-images of bright objects even when the latter have vanished from sight or indeed that even in total darkness we often see the most varied and often fantastic constructions without any corresponding objects, so likewise it is conceivable that during dreams the organ of consciousness develops an activity of fantasy quite independent of the outside world. A similar though attenuated kind of such activity in the form of imaginativeness is even useful and required for the formation of new combinations of ideas. Yet this activity too often overshoots the mark. A naive person regards the sun, moon, trees and well-springs as animate beings, but even one who is educated still imagines every force in the image of human exertion. In such cases we need strict checks and a sharp dismissal of everything that is mere dreamlike addition. This supervision in turn becomes a habit through constant practice. By pushing this habit to excess and applying it even where it does not belong, one arrives at the notion that all our ideas are dreams and nothing exists except the person that has them, that is one single dreamer. On Darwin's theory this aberration is just as understandable as the development of our normal activity of imagining. The mechanical nature of this activity has however been confirmed recently by the possibility of confusion even in a healthy state, through sleep, and even more in sickness, through hallucinations, feverish fantasies and madness.

From a Darwinian point of view we can grasp furthermore what is the relation of animal instinct to human intellect. The more perfect an animal, the more it shows incipient traces of intellect alongside instinct.

To an animal requiring only a small number of actions which moreover must constantly occur under highly similar conditions, it is of extreme usefulness if without much need for reflection it has an innate drive towards the right way of acting, as with birds, which without instruction are able by innate instinct to build nests with admirable skill. To us no doubt it would at first sight appear as a much more perfect state if without instruction and intense reflection we were able always to hit on the right measure. Whereas, however, under the simple conditions in which these animals live it was the easier and less complicated dispensation that the drive towards the entire mode of behaviour should be hereditary as a whole, this very feature stands in the way of any adaptation to changed circumstances, any progress; under complicated conditions of life man's innate ability is much superior for forming inner pictures of external events and thus collecting experiences according to which action can be regulated in each case.

Incidentally, in man instinct indeed recedes considerably, but its traces are still noticeable everywhere and by no means only in cases such as the above-mentioned sucking instinct, or the child's instinct of imitation, but also with all elementary instincts that suppress or anticipate thought in adults. Fright caused by sudden noise or fear by sudden danger involuntarily anticipate rational action in the same way as anger caused by sudden attack. The inherited habit of reacting violently to strong impressions, which is useful in giving our actions the required emphasis and liveliness, here exercises an invincible influence and becomes harmful if it runs too far ahead of reflection. Quite generally the basic drives of human character, craving for pleasure and inertia, but also ambition, thirst for power, sympathy and envy, all these arise from inherited dispositions, that is in the first place innate instincts. How far removed we are from pure rational grounds being the motives of all our actions! The innermost impulses to action mostly still arise from innate drives and passions, that is from instincts germinating within us without our concurrence, which do indeed become harmful and reprehensible if dominating the intellect, but nevertheless are necessary to lend our actions liveliness and our character its peculiar colouring. The machinery of the world maintains itself, as Schiller says, "today, as ever, by hunger and love, and the time is as yet far off when philosophy will hold the universal circuit together".

Superstition likewise is instinctive in character, and often some of the most educated people cannot quite rid themselves of it. It arises from the continued effect of our need for causality in cases where this is unjustified. The habit of looking for causal connections everywhere induces us to establish a causal link between events that seem purely accidental and with some other often disparate ones, so that the law of cause and effect which correctly applied is the basis of all cognition becomes a will o' the wisp that leads us on to quite erroneous paths.

We must further call to mind how well all the mechanism of our social arrangements too fits into the framework of these considerations. There we find countless rules of propriety and forms of politeness that seem, in part, so unnatural and forced that they seem absurd and ridiculous to that unprejudiced view often called reason but which in fact involves a forgetfulness of the omnipotence of mechanics. These rules of propriety are not the same at all times, and in the case of foreign peoples they often diverge so sharply from our own as to confuse us utterly, but exist they must.

The activity of conservative, pedantic, stiff and starchy arbiters of propriety who watch over the meticulous observation of every traditional custom and every rule of social intercourse and over the accurate use of all their titles in speeches as well as over the acknowledgement of all their social privileges often appears ridiculous to us; however, it is beneficial and necessary in order that social intercourse should not grow coarse. As to preventing formality from petrifying the life of the mind, this is looked after by the emancipated, unfettered and *hommes sans gêne*. The two types of people fight each other and together keep society in proper equilibrium.

In quite a different area of social life another mechanism is at work, maintaining equilibrium while there is constant and vigorous motion, one of the most grandiose and admirable mechanisms man has created, namely capital, money. We need only read Zola's novel *L'argent*. It has so much refined primitive barter as practised by ancient peoples, that the various forms of money mesh in with all the laws and traditional rules of commerce and stock exchange more admirably than the gears of the most complicated watch movement and they work with all the same liveliness, certainty and precision as the best designed electric motors.

Those who come off losers decry mammon; the swindler who twists the rules out of greed is expelled like useless matter from a living organism; but for our modern civilization money and stock exchange transactions are just as important as the art of printing, steam and electricity.

Does not the individual exercise magic power when a whole lot of metal pieces without intrinsic value become for him a means for creating palaces, parks, yachts, in short everything that embellishes life, and indeed for instituting prizes that long after his death continue to make an essential contribution to the creation of masterpieces in art and science? Yet the magician himself, is he not in his turn subject to the laws of mechanics, if the wrong position of a membrane in his heart or the rupture of a small blood vessel in the brain deprives him of the use of all this accumulated splendour, transforming the mighty one at a single stroke into a piece of dead matter?

Even contempt for paper money seems to me to be a one-sided attitude, for surely it has another side than just the one that Goethe has put into such a garish light in Faust. Indeed if we include all securities, bonds, bills of exchange and the like, it is almost the crowning element of the most important sector of human intercourse, namely of the mechanism that regulates what is *meum* and *tuum* in accordance with the complex requirements of our times.

To return from the sublime to the petty, let me point out that the irresistible instinct of cleanliness, which if neglected is soon revived by gossip, is highly useful because it removes all harmful infectious matter from our homes. Of course it overshoots the mark if for example brass surfaces are kept constantly polished, when their patina is not only harmless but positively soothing to the eye, given the glare of modern illumination at night. Still, I certainly do not want to maintain that we should be better off if dusting were allocated to bacteriologists rather than to domestics.

I should not find it hard to find further examples for my thesis, on the contrary, I should be in difficulties if I had to discover any process that was not an example.

Not only our bodily organs have thus been turned into a domain of mechanics, but also our mental life, even art and science, affective impressions and enthusiasm. Yet is not mechanics indeed rather too mechanical for the representation of all this? Take the most complicated mechanism made by human hand, how paltry and lifeless it is compared with the simplest vegetable or animal formation!

I foresee how the enthusiast will be horrified by these last remarks of mine, how he fears that everything great and noble is degraded to a dead and unfeeling mechanism and all poetry falls away. However, it seems to me that all this apprehension is based on a complete misunderstanding of what I said.

Indeed, our ideas of things are never identical with the nature of things. Ideas are only pictures or rather signs that necessarily represent one-sidedly and indeed can do no more than imitate certain modes of connection of the things signified without in the least touching their essential character.

We need therefore take back none of the sharpness and definiteness of our previous expressions. We have in any case used them only for asserting a certain analogy between mental phenomena and the simple mechanisms in nature. We have merely constructed a one-sided picture in order to illustrate certain connections between phenomena and to predict new ones thus far unknown to us. Alongside this one picture there can and because of its one-sidedness indeed must be others that represent the inward and ethical side of the matter; these latter will no longer hinder the exaltation of the soul, as soon as we take the right view of the mechanical picture which should be applied only where it belongs. Yet we shall not deny its usefulness but reflect that even the noblest ideas and conceptions are again only pictures or external signs for the ways in which phenomena are linked.

This does away with the objection that might perhaps have been raised against my remarks, that they ran counter to religion. Nothing is more perverse than linking religious concepts which rest on a quite different and immeasurably firmer basis with the vacillating subjective picture that we form of external things. I should be the last to put forward the views here mentioned, if they harboured any danger for religion. Yet I know that the time will come when all will own that they are as irrelevant to religion as the question whether the Earth is at rest or moving round the Sun.

While then the principle of mechanical explanation has conquered ever wider domains within the whole sphere of science, oddly enough it has lost some ground in its most central field, namely in theoretical physics. The cause for this, as often with victorious nations too, lay partly in internal divisions but partly also in external conditions.

While people most successfully aimed at working out applications of mechanics down to the smallest detail, there arose a line of approach that began to shake the basic pillars of the science, and pointed to obscurities in its principles. The fundamental concept of mechanics is that of motion. The concept of pure motion detached from any other change stands out in full clarity only in the study of rigid bodies, where indeed we have a perfectly immutable structure in which nothing changes save spatial position. Now there are in nature no perfectly rigid bodies but certainly solid ones that are subject to only imperceptible changes of shape during motion. As for changes in shape in liquids and gases, these one tries, without straining the facts, to reduce to the motions of their smallest constituent parts. Indeed to the eye they resemble changes of shape of a sand heap, which consists of individual perceptible grains. Nevertheless, in the case of actual fluids there is something hypothetical about the assumption that there too each individual particle is identifiable at all times, for experience shows that we are given only the constancy of total mass and weight.

It was now attempted to prove a priori that every change, even if apparently qualitative, must be reducible to a motion of the smallest parts, motion being the only process in which the object moved remains always the same. All such metaphysical reasons seem to me to be insufficient. Of course we cannot avoid forming the concept of motion. If therefore all apparently qualitative changes were representable by the picture of motions or changes of arrangement of smallest parts, this would lead to an especially simple explanation of nature. In that case nature would appear to us at its most comprehensible, but we cannot compel her to this, we must leave open a possibility that this will not do and that we need in addition other pictures of other changes; understandably, it is precisely the more recent developments of physics that have made it prudent to allow for this possibility.

Mechanist physics had conceived all bodies as aggregates of material points acting directly on each other at a distance. At very small molecular distances it was to be cohesive, adhesive and chemical forces that were acting, and at larger distances gravitational ones. Besides ponderable matter one assumed that there was a luminous aether conceived as just like a solid body, whereas electro-magnetic phenomena, as mentioned earlier, were explained by means of electric and magnetic fluids whose parts again acted directly on each other at a distance. This last hypothesis was long able to account for all observed phenomena. Only a little over ten years ago did Hertz succeed in giving experimental proof that, as Faraday and Maxwell had suspected already, electric and magnetic forces do not act directly at a distance but are caused by changes of state that are propagated from one volume element to the next at the speed of light. This gave the revered old theory of electric and magnetic fluids a blow from which it soon expired, but another theory too was damaged by Hertz's experiments. For the laws of propagation of electro-magnetic waves showed such absolute agreement with those of the motion of light, that no further doubts could be entertained as to the identity of the two phenomena. If this did not as yet refute the theory that light rests on the vibratory motion of the smallest particle of a luminous aether, nevertheless it had been established that this aether certainly must have different and much more complex properties than had hitherto been ascribed to it. This gave such preponderance to the theory of electricity and magnetism that from some quarters it was attempted to replace the mechanical hegemony in theoretical physics by an electro-magnetic one, by trying conversely to derive the simplest laws of mechanics from the theory of electro-magnetism.

On the other hand one had become suspicious of all hypotheses and confined the task of theory to supplying a description of phenomena without anywhere going beyond what was given in experience. This leaves the choice between two extreme methods. If one frames hypotheses that are too specialized, one runs the danger of introducing into one's field of ideas items that are superfluous or even incorrect. If, however, one tries to ward off all hypothesis, theory becomes vague and unsuitable for predicting entirely new phenomena thus guiding experimentation into new pathways. It is understandable that a period of excessively bold hypothesis should have been followed by a corresponding reaction.

Moreover a certain concept, already clearly recognized by Leibniz as important and long since playing a significant role in mechanics, gradually grew into a mighty bond encompassing the whole world of phenomena, namely the concept of energy. Although more abstract than the concept of matter, it could be accurately followed and even quantitatively determined in those phenomena where we lack all indications of matter to which it might be tied.

In every one of its manifestations energy shows different characteristic peculiarities but also remarkable analogies, so that the theory of the properties and transformation of energy soon become so influential as in its turn to seek hegemony in theoretical physics, attempting to make of it the science of energetics. There is no need to discuss this further here, since both the most extreme theories of action at a distance and energetics have been so clearly dealt with by two members of this university in their respective inaugural lectures[1].

As regards the formal logical foundation, traditional mechanics had adopted the dualism of force and matter. Matter is that which is movable. Now we are in the habit of looking for the cause of every special motion. By extending this habit beyond the boundaries of its justifiable domain thus overshooting the mark in applying it, one thought that it was necessary to assume a cause separate from matter for the fact that there are any phenomena of motion at all, which one then called force, attributing to it separate existence alongside matter. Kirchhoff denied that this was necessary and thought he could manage by merely assuming matter and the fact that it moved according to certain laws to be described, but he retained direct action at a distance. However, if we seriously ask what is left of this action in the light of our present views, we no longer find a great deal.

Electric and magnetic forces do not act at a distance but from one volume element to the next. Nor can we establish action at a distance in the case of elastic and chemical forces or of adhesion and cohesion, whose effective range is in any case minute. Only gravitation is left, but there too the analogy with electrostatic and magnetic forces makes it very likely that there is transmission through a medium.

Although Newton called even direct action at a distance a mere expedient, nevertheless the whole edifice of classical mechanics is cut to this pattern. It is therefore not surprising that Hertz tried to reform mechanics from its foundations, replacing accelerating action by equations of consstraint. Yet Hertz too constructed matter from material points, which indeed do not exert forces at a distance on each other, but which are just as immediately linked at a distance by the constraints. It is as though Hertz replaced action at a distance by constraints at a distance.

Brill has tried to apply Hertz's method to continua, and was able in this way to derive the equations of motion for incompressible fluids. With Lord Kelvin one might now explain nature from the mutual interplay of vortex rings or other phenomena of motion in such fluids, which might even have rigid structures immersed in it. This would indeed amount to a picture of the entire world of phenomena on the basis of Hertzian mechanics. However, we see at once that it would not differ greatly from the old fantastic world pictures. The gain would be by no means so great as promised by the beautiful philosophic foundation of Hertzian mechanics. So far, however, it has not been possible to develop the latter in some other way that is less encumbered by hypotheses.

Whereas the most recent views of electro-magnetism merely try to seek salvation in the action of volume elements on their neighbours, some very recently observed phenomena in cathode rays and electrolysis have led to the assumption that even electricity has an atomic constitution, consisting of discrete elements, namely electrons. We therefore see that the old Kantian antinomy opposing infinite divisibility of matter to atomistic constitution still keeps science in suspense, except that today we do not regard the two views as infected by internal logical inconsistencies arising from the laws of thought, but we take each as a mental picture we have constructed and we ask which can be developed more clearly and more easily while most correctly and definitely reproducing the laws of phenomena.

Summing up in conclusion, our result is that one side of all processes of inanimate and animate nature is representable, or, as the phrase goes, made intelligible with a measure of exactness not hitherto achieved in any other way, while at the same time none of the higher endeavours and ideals are in the least impaired.

Finally, a word to you, my future pupils and student colleagues: be full of idealism and enthusiasm in absorbing everything that your alma mater offers you, but in assimilating it be mechanical and go on working untiringly and uniformly like machines.

II. VIENNA, OCTOBER 1902

It is customary in inaugural lectures to begin with a hymn of praise to one's predecessor. Today I can save myself this sometimes troublesome task, for even if Napoleon I never managed to be his own great-grand-father, I on the other hand am currently my own predecessor. I can there-fore embark straight away on the treatment of my real topic.

As to that, the giving of inaugural lectures on the principles of mechanics is something for which I have by now acquired a certain amount of prac-tice. The lecture with which I began my tenure as ordinary professor at the university of Graz 33 years ago already dealt with this topic. Since then this is the third occasion on which I open my lectures in Vienna by con-sidering this matter, and in addition there is one inaugural lecture in Munich and one in Leipzig, both on the same subject.

It is indeed important enough to stand repeated treatment without too much risk of repetition. Mechanics is the foundation on which the whole edifice of theoretical physics is built, the root from which all other branches of science spring. This becomes clear if on the one hand we consider the historical development of the physical sciences and on the other examine their inner logical connection.

However much science prides itself on the ideal character of its goal, looking down somewhat contemptuously on technology and practice, it cannot be denied that it took its rise from a striving for satisfaction of purely practical needs. Besides the victorious campaign of contemporary natural science would never have been so incomparably brilliant, had not science found in technologists such capable pioneers.

To find the first traces of man's mechanical activity we must transplant ourselves from the present age of X-rays and wireless telegraphy back to the very earliest beginnings of human culture. The first human tool was the cudgel. The orang-utang wields it too, and for a purpose to which even today, when we feel so superior to that animal, a goodly portion of human inventiveness and technical ingenuity is addressed. What shall I call this purpose? The friends of peace call it murder, soldiers call it risking the supreme sacrifice of life itself for mankind's noblest goods, for honour, freedom and country.

However this may be, we must regard the cudgel as a first mechanical tool, the first gift of an incipient sense for technology. When later the civilization of mankind began to develop, it was not acoustic or optical apparatus, thermal or even electro-magnetic machines that were invented first. Things went somewhat more slowly. The need for better sealing of caves and making artificial ones gradually led to the building of shelters and castles. Since for this purpose it was necessary to fetch heavy stones or mighty tree-trunks, inventiveness was stimulated. Man rounded suitably shaped branches into rollers and later built roughly worked wheels; he used the cudgel as a lever in the most primitive form and thus entered the field of mechanics in the narrower sense, unconsciously at first, but afterwards with increasingly deliberate consciousness.

Hats off to these inventors in bearskins and bark shoes. The man who first used deftly inserted rollers to move a stone whose weight seemed forever to defy the giant fists of his fellows surely experienced no less satisfaction than Marconi on perceiving the first airborne trans-oceanic telegraphic signal, provided of course that what the papers report about it is wholly true.

From such unspectacular beginnings mechanics grew up, slowly at first, but steadily and later at ever increasing pace. What had been achieved by his time moved Archimedes to such admiration that he would have ventured to lift the Earth from its foundations if only there had been a firm fulcrum for him. Today's technological progress may not have moved the Earth, but the whole social order and the whole mode of human intercourse has indeed almost been lifted from its hinges.

Indeed, progress in the field of the natural sciences has fundamentally transformed man's whole mode of thinking and feeling. Whereas the earlier humanist age saw everything as animate and sensitive, we are regrettably becoming more and more inclined to view everything under the aspect of a machine. In the past the rambler roamed singly through woods and fields while in a coach you could do nothing better than dream and indulge poetic fancy, unless vexation happened to outweigh boredom; today we work and calculate even in express train or ocean liner. In the past the coachman used to direct his horse's will by encouraging it in human language, today we operate electric motors or crank up our motor car in silence.

And yet we cannot rid ourselves of the idea that nature is animate. Are not today's big machines working like conscious beings? They snort and puff and howl and moan, uttering sounds of complaint, fear and warning and whistling shrilly under excess power. They absorb from their surroundings the materials needed to maintain their power, eliminating what is unusable, subject to exactly the same laws as our own bodies.

I find it peculiarly attractive to imagine how happy the pathbreaking geniuses of the most varied fields would be at the achievements of their successors who in many ways stood on their shoulders; for example, what would Mozart feel today if he could hear a master performance of the Ninth Symphony or of Parsifal. The great Greek philosophers of nature and above all that ardent mathematical genius, Archimedes, would have said roughly the same to the achievement of present-day technology, they would certainly not lack a sense and fervour for the sublime. Even today the highest degree of fervour is denoted by the beautiful Greek word 'enthusiasm'.

But I seem to digress somewhat; let us return to our topic.

So far I have spoken all the time about machines and technology. However, you would go seriously astray if you expect that the aim of my lectures will be to initiate you into the art of building machines; that is the task of technical mechanics and the theory of machines, whereas my topic will be analytical mechanics, whose definition is much more general. It must explore the laws according to which the totality of phenomena of motion in circumambient nature occurs.

To begin with we there find very many bodies that have constant shape at least as for as observation goes. Their motion is thus merely change of position and rotation without any change of shape; analytical mechanics will for a start have to indicate the laws for such changes of position. Other bodies, fluids (liquids and gases) undergo constant and most varied changes of shape during motion. We can form an illustrative picture of this unceasing change of shape by conceiving of fluids as made up of smallest particles each of which moves independently according to the same laws as solid bodies, but always in such a way that two neighbouring particles of the fluid always move in approximately the same way. In addition to the forces acting externally on each particle we must consider those exerted by the various particles on each other. In this way the motion of fluids can be reduced to the laws of mechanics for solid bodies.

Phenomena of motion are the ones that we observe most often and most directly, all other natural phenomena are more concealed. Besides we can cope with phenomena of motion by means of the least number of concepts, all we need for describing them are the concepts of position in space and its change in time, whereas for other phenomena many less clear concepts are required, such as temperature, luminous intensity and colour, electric tension and so on.

It is the ubiquitous task of science to explain the more complex in terms of the simpler; or, if preferred, to represent the complex by means of clear pictures borrowed from the sphere of the simpler phenomena. Therefore in physics too the attempt was made to reduce all other phenomena, such as sound, light, heat, magnetism and electricity, to mere motions of the smallest particles of these bodies, and this turns out quite successful for very many phenomena, although of course not for all. Thus the science of motion, that is mechanics, became the root of the other physical disciplines, which seemed increasingly to grow into special chapters of mechanics.

Only in more recent time has there been a reaction to this. The difficulties occasioned by a purely mechanical account of magnetism and electricity raised doubts as to universal mechanical explicability and it was precisely electro-magnetism that became increasingly important not only in practice but also for theory. In the end it became so powerful as to attempt a reversal of roles, explaining mechanics in terms of electro-magnetism. Whereas previously one had tried to explain magnetism and electricity by rotations or oscillations of the smallest parts of bodies, the present aim was to derive the fundamental laws of motion for bodies themselves from the laws of electro-magnetism.

The best known law of mechanics is the principle of inertia. Every high-school boy is familiar with it today – I speak of course only of inertia in the physical sense. Until recently the law of inertia has been regarded as the foremost fundamental law of nature, itself inexplicable but essential for explaining any phenomenon. However it is a consequence of Maxwell's equations for electro-magnetism, that a moving electric particle without any mass or inertia of its own must move, by the mere action of the surrounding aether, just as though it had inertial mass. This led to the hypothesis that bodies have no inertial mass, but only mass-less electric particles or electrons, their inertia being a mere semblance provoked by the surrounding aether while they move through it. Similarly it was possible to reduce the action of mechanical forces to electro-magnetic phenomena. Whereas previously one wanted to explain all phenomena in terms of the action of mechanisms, now it is the aether that is a mechanism, quite obscure in itself of course, that is to explain the action of all other mechanisms. It was no longer a question of explaining everything mechanically, but of finding a mechanism to explain all mechanisms.

What, then, is meant by having perfectly correct understanding of a mechanism? Everybody knows that the practical criterion for this consists in being able to handle it correctly. However, I go further and assert that this is the only tenable definition of understanding a mechanism. Of course one might object that it is conceivable for a person to have learnt the way to handle a mechanism without understanding the mechanism itself, but this will not hold water. The reason for our saying that he does not understand the mechanism is merely that his knowledge of how to handle it is confined to its regular operation: as soon as something is broken or malfunctioning or some other unforeseen disturbance occurs, he no longer knows what to do. On the other hand we say of somebody that he understands the mechanism if he knows the right thing to do in all these cases as well. This circumstance therefore really does seem to constitute the definition of understanding. How we are to form our concepts cannot be defined and is indeed quite indifferent, so long as they always lead to the correct mode of action.

Thus solipsism rests on a well-known and tempting mistaken inference. Solipsism is the view that the world is not real, but a mere product of our fantasy, like a dream object. I too once hankered after this whim, which led to my failing to take the right practical action and caused me damage; to my immense delight, for this provided me with the desired proof that the external world exists, a proof that can consist only in showing that if we doubt this existence we are less able to act appropriately.

When thirty-three years ago I gave my inaugural lecture on mechanics mentioned above, one of my colleagues in Graz teased me by saying: "How can you spend your time on something so purely mechanical?" He naturally meant no more than a play on words, but I reacted and tried to show that mechanics was nothing mechanical. Yet in spite of its difficulty, in spite of the infinite effort in ingenuity spent on its development by the best intellects through centuries, the mechanical character of our science still makes itself felt.

I have already discussed the concept of inertia. A second fundamental concept of mechanics is work. The most important law of mechanics might be put as follows: nature does everything with a minimum expenditure of work. Who would not find trivial side-issues occurring to him when hearing this! Is not the concept of work one of the most important and at the same time one of the most enigmatic concepts for practical life as well as for natural science as a whole? The first human pair expelled from Paradise already saw work as the supreme curse, yet without work man would not be man. Constant and unremitting work is something that man indeed shares with beasts of burden and even with inanimate machines that he himself has made, but nevertheless industriousness is praised as one of the best features of any man's character, be he ruler or day labourer.

In conclusion let me raise the question whether mankind has become happier through all this progress of culture and technology, indeed, a delicate question. Certainly nobody has yet invented a mechanism for making men happy, happiness is something that everybody must seek and find within himself.

However, science and civilisation have succeeded in eliminating influences that disturb happiness, by managing in many cases to fight off the dangers of lightning, epidemics of whole populations and diseases of individuals. Besides, science provides greater opportunities for finding happiness by giving us the means of roaming our beautiful Earth more easily and coming to know it better, of visualizing the structure of the celestial sphere more vividly and surmising at least obscurely the eternal laws of nature as a whole. In this way it enables mankind to develop its bodily and mental forces even further, giving us increasing sway over all the rest of nature and allowing the person who has found inner peace to enjoy that state more perfectly by enhanced development of his powers.

It is my task in the present course of lectures to offer you many things: intricate theorems, ultra-refined concepts and complicated proofs. Please forgive me if today I have as yet done little of this. I have not even, as would have been proper, defined the concept of my science, namely theoretical physics, nor yet developed the plan according to which I intend to treat this topic in these lectures. Today I did not want to present all this to you, I think that later in the course of our work we shall be better able to clarify these things. Today I merely wanted to present something more trifling, although for me it happens to be all I have, namely myself and my whole way of thinking and feeling.

Likewise during these lectures I shall have to demand many things from you: concentrated attention, unremitting diligence, inexhaustible will-power. Forgive me therefore if before embarking on any of this I ask you in return for something that means most to me, namely your confidence, affection and love, in a word, the most precious thing you can give, namely yourselves.

NOTE

[1] *Editor's note*: The allusion is presumably to the inaugural lectures of J. C. F. Zöllner, *Über die universelle Bedeutung der mechanischen Prinzipien* (Leipzig 1867), and W. Ostwald, *Die Energie und ihre Wandlungen* (Leipzig 1887).

AN INAUGURAL LECTURE
ON NATURAL PHILOSOPHY*

You have come here in unusually large numbers to listen to the modest inaugural remarks that I have to address to you today.[1] I can explain this to myself only from the fact that my current lectures are indeed in some respect a curiosity in academic life, not so much by their content or form, but by the attendant subsidiary circumstances.

For I have hitherto written only one single dissertation of philosophic content and what moved me to do so was pure chance. On one occasion in the assembly hall of the academy, I was involved in the liveliest debate with a group of academics, Professor Mach amongst them, on the newly revived controversy about the value of atomistic theories.

I here mention in passing that in the assignment that begins with today's lecture I am in a certain sense Mach's successor and I should really have started my talk by paying homage to him. However, I think that to express his special praise would amount to carrying owls to Athens, as far as you are concerned and indeed not only you but any Austrian and even any educated person throughout the world.

Mach himself has ingeniously discussed the fact that no theory is absolutely true, and equally hardly any absolutely false either, but that each must gradually be perfected, as organisms must according to Darwin's theory. By being strongly attacked, a theory can gradually shed inappropriate elements while the appropriate residue remains. I therefore think that the best way to honour Mach is in this sense to contribute as far as within me lies to the further development of his ideas.

In that group of academics during the debate on atomism Mach suddenly said laconically: "I do not believe that atoms exist". This utterance ran in my mind.

It was clear to me that we unite groups of perceptions into ideas of objects, for example of a table, a dog, a man and so on. Moreover we have memory pictures of these groups of ideas. When we form new groups of

---

* *Populäre Schriften*, Essay 18. First published in *Die Zeit*, 11 December 1903. See Note (1).

ideas which are quite similar to these memory pictures, there is sense in the question whether the corresponding objects have existence or not. We here have as it were an accurate measure for the concept of existence. We know exactly what is meant by the question whether the griffin, the unicorn or a brother of mine exists. However, when we form quite novel ideas, such as those of space, time, atoms, the soul, or even God, does one know, so I asked myself, what is meant by asking whether these things exist? Is not the only correct thing to do here to try to clarify what concepts one is linking with the question as to the existence of these things?

Discussions of this kind formed the topic of my one and only dissertation in the field of philosophy. As you see, it was genuinely philosophical; abstruse enough, at least, to deserve the name. Apart from this I have published nothing in this field. This much might of course pass; if one wanted to be malicious one might say that here and there people have taught at universities who had written even one publishable piece less about their fields.

In any case, however, it must fill me with utter modesty. It is said that if God gives one a task, he will give one the wit to do it. Not so the ministry: it can of course make appointments and fix salaries but it can never furnish the wit; for that I alone must bear responsibility.

Not only while writing my only dissertation but at other times too, I have often speculated about the enormous field of philosophy. It seems infinite and my powers slight. A whole life would be but little time to struggle through to some results in it; the tireless activity of a teacher from youth to old age would not suffice to transmit philosophy to the next generation, am I then to pursue it as a subsidiary occupation along with another subject that by itself requires all my powers?

Schiller has said that a man grows with his purposes. Dear old Schiller! I fear that man does *not* grow with his higher purposes.

When I had qualms about taking on this heavy burden, I was told that another would do no better than I. How paltry this consolation seems at the moment when I must shoulder the load.

And yet, what bows me down now could surely raise me up again? If I, who have busied myself so little with philosophy, was found to be the most deserving person to lecture on it, is this not a twofold honour?

If it is desirable for a professor of medicine or engineering that he should continue in practice alongside his teaching, lest he become ossified, indeed if Moltke was made a member of the historical division of the Berlin Academy because he had not written but made history, perhaps I too was chosen not because I had written about logic, but because I belong to a science that offers the best opportunities for daily practice in strict logic.

If it was only with hesitation that I followed the call to meddle with philosophy, philosophers have the more often meddled with natural science. They have now for many years invaded my preserves and I could not even understand what their views were and therefore wanted to improve my knowledge of the fundamental theories of all philosophy.

To go straight to the deepest depths, I went for Hegel; what unclear thoughtless flow of words I was to find there! My unlucky star led me from Hegel to Schopenhauer. In the preface to his first work I found the following passage which I will report verbatim here: "German philosophy stands laden with contempt, derided by other countries, banished from honest science like a" ... I suppress the next bit since there are ladies present. "...The heads of the present generation of learned men are disorganized by Hegelian nonsense. Incapable of thought, uncouth and stupefied, they fall prey to shallow materialism that has crawled forth from the basilisk's egg". With that I was of course in agreement, except that I found that Schopenhauer, too, really deserved the blows of his own cudgel.

However, Herbart's calculations about psychological phenomena equally seemed like a travesty of analogous calculations in the exact sciences. Even in Kant there were many things that I could grasp so little that given his general acuity of mind I almost suspected that he was pulling the reader's leg or was even an impostor. In this way there developed in me an aversion from philosophy at that time, indeed a hatred of it. In view of these old philosophical systems I should almost like to say that in my case the goat has been made the gardener. Or was I given precisely this teaching appointment in the way that an old democrat is given the title of Court Councillor, in order to turn Saul definitively into Paul? I am afraid that in these lectures I shall be oscillating between goat and councillor, and although I hope that I shall never fall into the style of which I have just read you a sample, I may nevertheless be somewhat rough in places when I apply Mach's method to the perfecting of philosophical systems.

My aversion from philosophy at the time was incidentally shared by most natural scientists. Every metaphysical orientation was pursued with the aim of extirpating it root and branch; but this attitude did not last. Metaphysics seems to cast an irresistible spell on the human mind and all the abortive attempts at lifting its veil have not impaired its power. The drive towards philosophizing seems ineradicably innate. Not only Robert Mayer, who was a philosopher through and through, but Maxwell, Helmholtz, Kirchhoff, Ostwald and many others too made willing sacrifice to it and recognized its questions as the very highest, so that today metaphysics figures once again as the queen of sciences.

A man like Francis Bacon*, who stood at the cradle of inductive science, already called it a virgin consecrated by God; although he adds maliciously that just because of this noble property it must remain forever barren. And, to be sure, many investigations in the field of metaphysics have remained barren. Let us nevertheless try out whether all speculation must really be so. As soon as we start our activity, we find the great difficulty of fixing the concept of philosophy. [Here the lecturer goes through the main definitions of philosophy used to date and all of them appear to him untenable. Then he continues.] In such difficult matters it is in the first place important to formulate the question correctly. It can be framed in the following forms: (1) How has philosophy been defined by different philosophers? (2) Which definition would correspond most closely to ordinary usage? (3) Which seems the most appropriate? (4) How, regardless of what others have done, what corresponds to usage and what is appropriate, would irresistible compulsion lead me to frame the concept of philosophy? How do my inner sense and every strand of my thinking force me to resolve the question? Each of these questions can be split and analysed into several others. Even then we should not have attained absolute thoroughness, but let us not pursue the analysis, because we now seem to understand each other passably well.

I will now answer the question in this last sense: which definition forces itself on me with irresistible compulsion? Here I always had a nightmarish feeling that it was an unresolvable puzzle how I could exist at all, or that a world could, and why it should be precisely as it was rather than otherwise. The science that would succeed in resolving this puzzle seemed to me the greatest and true queen of sciences, and this I called philosophy.

* *Editor's note:* By some confusion, Boltzmann attributes this remark to 'Roger Bacon of Verulam'.

I gained more and more knowledge of nature, I absorbed Darwin's theory and saw from it that it was really a mistake to ask in such a way that the question cannot be answered, but the question always returned with the same compelling violence. If it is unjustified, why can it not be dismissed? Following on from this there are countless other questions: if there is something else behind perceptions, how can we even suspect that there is?[2] If there is nothing behind them, would then a landscape on Mars or on a planet of Sirius really not exist if no living being is ever able to perceive them? If all these questions are senseless, why can we not dismiss them or what must we do in order finally to silence them? It is to be the task of my present lectures at least to seek to clarify these questions.

So far I have no idea where such illumination is to be found and therefore I live in a truly Faustian mood. Indeed it is Faust who says: "I am to teach with bitter sweat what I do not know myself". Nor would I teach it, but merely collect together everything that might contribute slowly to bringing light into this darkness and to inspiring you to collaborate with me as best you can to further the attainment of this goal.

My method of lecturing may seem strange to some, but perhaps it is really and truly academic. In the best sense of the word, academic presentation aims less at teaching ready-made solutions of problems but rather at posing problems and giving hints towards their solution. We shall therefore go through the various fundamental concepts of all sciences and examine them all with regard to the goal we have set ourselves, sub specie philosophandi.

The very title I have given to my present lectures is a stumbling-block, for it is the literal translation of the title of the first and greatest work ever written about theoretical physics, namely the *Principia Philosophiae Naturalis* of Newton. If I understood it in the same sense as Newton, I would have to present an outline of theoretical physics. I chose the title only to show you how little philosophy can cleave to words: these are exactly the same, but today we understand something totally different by them from what Newton did then, and some conservative Englishmen still do today.

Let me now hurry to my conclusion. I concluded my first lecture in Vienna in a way that pleased me especially, not because of content or form but because what I said expressed exactly what I felt; not because it was cleverly contrived but because it was not contrived. Today I feel exactly the same and therefore cannot but express myself by using the self-same words. I then said*:

"It is my task in the present course of lectures to offer you many things: intricate theorems, ultra-refined concepts and complicated proofs. Please forgive me if today I have as yet done little of this. I have not even, as would have been proper, defined the concept of my science, namely theoretical physics, nor yet developed the plan according to which I intend to treat this topic in these lectures. Today I did not want to present all this to you, I think that later in the course of our work we shall be better able to clarify these things. Today I merely wanted to present something more trifling, although for me it happens, to be all I have, namely myself and my whole way of thinking and feeling.
Likewise during these lectures I shall have to demand many things from you: concentrated attention, unremitting diligence, inexhaustible will-power. Forgive me therefore if before embarking on any of this I ask you in return for something that means most to me, namely your confidence, affection and love, in a word, the most precious thing you can give, namely yourselves."

Let these words again conclude my talk to you today.

* *Editor's note*: The following paragraphs, drawn from Boltzmann's previous inaugural lecture, appear also on p. 152 above.

NOTES

[1] Since quite erroneous views about my first lecture on natural philosophy (given on 26 October) have evidently been spread abroad, partly because of defective press reports, I am glad to follow the request of the editor of *Zeit* to publish the lecture. Since it was delivered quite extempore I cannot vouch for the precise words though certainly for the sense.
[2] The need, alongside sensations, for an instinct to think objects was pointed out, if I understand him aright, by Schopenhauer, of whom we have just spoken so ill.

ON STATISTICAL MECHANICS*

My present talk has been ranged under 'applied mathematics' whereas my work in teaching and research is devoted to the science of physics. The great rift that has torn that science into two camps has hardly been more sharply defined than in the organisation of the talks to be presented at this scientific congress, which had to get through such enormously extensive material that it may be described as a flood, or to preserve local colour, as a Niagara of scientific talks. I am referring to the division into theoretical and experimental physics. While as a representative of theoretical physics I have been put into Section A for normative science, experimental physics turns up only much later under Section C for physical science. In between there are history, linguistics, literature, theory of art and theory of religion. Across all this the theoretical physicist must reach out his hand to his experimental colleague. We shall therefore not be able completely to avoid the question whether it is justified to divide science in general into two portions, and physics in particular into theoretical and experimental.

Let us consult first of all an investigator from a period when science had hardly grown beyond its first beginnings, namely Immanuel Kant. He demanded of each science that it should be developed from unitary principles and well-knit theories in a strictly logical manner. Natural science he regarded as hardly a full science except insofar as it is built on a mathematical basis. Thus chemistry in his time he did not count among the sciences, because it rested on merely empirical foundations and lacked a unitary regulative principle.

Viewed from this position theoretical physics would stand in advantageous opposition to experimental physics, occupying a higher rank as it were. Experimental physics would merely have to bring together the

* *Populäre Schriften*, Essay 19. Address given to the Scientific Congress in St. Louis, 1904, and first published in a translation by S. Epsteen as 'The Relations of Applied Mathematics' in *Congress of Arts and Sciences... St. Louis 1904*, Boston, 1905. The present translation is by Dr Paul Foulkes.

building bricks whereas it would be the task of the theoretical branch to make a building from them.

However the ranking order turns out differently if we consider the achievement of the last few decades as well as the progress that may be expected for the near future. The chain of experimental discoveries of the previous century was fittingly closed by the discovery of Röntgen rays. Following on from these our present century has discovered a veritable horn of plenty of new radiations with the most puzzling properties deeply affecting our whole view of nature. The revelation of such entirely new facts promises the greater future results the more puzzling and opposed to traditional views everything seems at the start. However it is not my task to discuss these experimental findings here. Rather, I must leave the gratifying task of descibing the fruits that have so to speak been garnered daily in this field and those that are still to be expected to those who represent experimental physics at this congress.

The representative of theoretical physics is by no means in an equally happy position. In his field too there is currently much activity. One might almost say that it is in the course of a revolution. Yet how intangible these results are when compared with the experimental ones. It here becomes evident that in a certain sense experiment is entitled to precedence over all theory. An immediate fact is intelligible at once and its fruits can supervene in a very short time as for example the various applications of Röntgen rays, the use of Hertzian waves for wireless telegraphy. The conflict of theories, however, is an infinitely lengthy business, indeed it almost seems as though certain controversies as old as science itself will live on as long as it does.

Every securely ascertained fact remains for ever immutable, at most it can be extended or complemented by the arrival of new items, but it cannot be entirely overthrown. This explains why the development of experimental physics proceeds continuously without any leaps that are too sudden and why it is never visited by great revolutions or commotions. It is very rare for something to be regarded as a fact and afterwards found to have been erroneous and even when it does happen the error will soon be cleared up without this greatly affecting the edifice of science as a whole.

It is of course strongly emphasized that any inferred truth that is recognized as logically necessary must continue to exist incontrovertibly. Yet even if we can hardly doubt this, experience nevertheless shows that our theoretical edifices are by no means built out of only such logically incontrovertibly grounded truths. On the contrary, these edifices consist of often arbitrary pictures of the connection between phenomena, that is of hypotheses.

Without somewhat, even if slightly, going beyond what is directly perceived, there is no theory nor even a perspicuously comprehensive description of natural facts fit to serve the prediction of future phenomena. This holds as much of the old theories whose territory is currently being challenged as of the most modern ones which suffer under a great illusion if they consider themselves free from hypotheses.

Of course we can keep our hypotheses fairly indefinite or even frame them in the form of mathematical formulae or in words that express a thought equivalent to them. Then we can check step by step that there is agreement with what is given; a total overthrow of what was previously built is of course not to be excluded even then, if for example the law of conservation of energy turned out at last to be false after all. However such a revolution will be extremely rare and in some cases so improbable as to be inconceivable.

A theory that is kept thus indefinite and unspecific may well serve as a valuable guide in experiments conducted along paths already cleared and for a detailed working out of knowledge already gained, but beyond this it is of no use.

In contrast with this, hypotheses that leave some play to fantasy and go more boldly beyond what is given will give constant inspiration for novel experiments and thus become pathfinders to totally unsuspected discoveries. Such a theory will of course be subject to change and it may happen that a complicated theoretical structure will collapse and be replaced by a new and more effective one, in which however the old theory as a picture of a restricted field of phenomena usually continues to find a place within the framework of the new one; as for instance emission theory for describing catoptric and dioptric phenomena, the hypothesis of an elastic luminous aether for representing interference and diffraction, the theory of electric fluids for describing electrostatic phenomena.

As to powerful revolutions, it would seem that they are likely to affect even theories that proudly describe themselves as free from hypotheses. No one, for example, will doubt that the theory known by the name of energetics must radically change its garb if it is to go on existing.

It has been urged against physical hypotheses that they have at times proved harmful and a hindrance to scientific progress. This reproach is based mainly on the role that the hypothesis of electric fluids had played in the development of electric theory. This hypothesis was brought to a high pitch of perfection by Wilhelm Weber and the universal recognition that his work has gained in Germany was there, indeed, an obstacle to the study of Maxwell's theory, somewhat as Newton's theory of emanation had stood in the way of wave theory. Such inconveniences will no doubt not be entirely avoidable in future either. One will always strive to give the currently prevailing view its most complete and perfect form. If now such a self-consistent theory never meets any resistance from experience, it does not matter whether it consists of mechanical pictures, geometrical illustrations or a body of mathematical formulae. It will always be possible that a new and so far untested theory will arise which represents a much wider but so far unknown field of phenomena. In that case the old theory will retain the greater number of supporters until the new field of phenomena is made accessible to experiment and crucial tests put the new theory's superiority beyond doubt. It is certainly useful to set up Weber's theory as a warning example for all time that we should always preserve the necessary mental flexibility. But this certainly does not diminish the merits of Weber of whose theory even Maxwell speaks with the greatest admiration. Nor can this case be advanced against the utility of hypotheses, since Maxwell's theory was to start with no less full of hypothetical assumptions than any other and only after it had been generally acknowledged was it freed from such elements by Hertz, Poynting and others.

The opponents of hypotheses in physics moreover voiced the reproach that the creation and further development of various mathematical methods for calculating hypothetical molecular motions had been useless or even harmful. This reproach I cannot accept as justified. If it were, the choice of my present topic would have to be termed equally mistaken and let this circumstance be my excuse for my having here once more enlarged on the much-canvassed topic of the use of hypotheses in physics, seeking to justify such use.

For the topic I have chosen for today's talk is not the whole development of physical theory. Some years ago I dealt with this topic at the German conference of natural scientists in Munich* and although quite a few things

* *Editor's note*: See pp. 77–100 above.

have happened since then I should nevertheless have had to repeat myself in many matters now. Moreover one who belongs decidedly to a definite party is less able to assess the other parties quite objectively. I am not now talking of criticising their value, my talk is not meant to criticise but merely to report. Besides I am convinced that the views of my opponents are valuable and I rise in defence only when they wish to belittle the value of mine. Yet giving a perfectly objective account, an uncovering of the mutual engagement of all the intellectual threads is just as difficult with another's view as with one's own.

Therefore let me choose as goal of the present talk not just kinetic molecular theory but a largely specialized branch of it. Far from wishing to deny that this contains hypothetical elements, I must declare that branch to be a picture that boldly transcends pure facts of observation, and yet I regard it as not unworthy of discussion at this point; a measure of my confidence in the utility of the hypotheses as soon as they throw new light on certain peculiar features of observed facts, representing their interrelation with a clarity unattainable by other means. Of course we shall always have to remember that we are dealing with hypotheses capable and needful of constant further development and to be abandoned only when all the relations they represent can be understood even more clearly in some other way.

Amongst the earlier mentioned questions as old as science itself but hitherto unsolved there belongs the query whether matter is to be conceived as continuous or as composed of discrete constituents (of very many but not mathematically speaking infinitely many individuals). This is one of the difficult questions that make up the border region between philosophy and physics.

As recently as a few decades ago scientists were very diffident about plunging into discussions on such questions. This precise question is of too much topical interest to science to be completely ignored, but we cannot discuss it without at the same time touching on even deeper problems such as the nature of causality, matter, force and so on. It is these last of which one used to say that they were of no concern to the scientist and should be handed over entirely to philosophy. Today things have changed considerably, indeed scientists are now evincing strong predilections for discussing philosophical topics and indeed rightly so. One of the first rules of natural inquiry is never to bestow blind trust on the instrument with which one works, but to scrutinize it in all directions. How then could we commit ourselves blindly to the concepts and views that are innate to us or have developed in the course of history, the more so since we have already enough examples where they led us astray. However, once we examine the simplest elements, where would be the boundary between science and philosophy at which we could stop?

I hope that none of the philosophers present will resent it or construe it as a reproach if I frankly say that assigning these questions to philosophy has perhaps produced bad results too. Philosophy has been remarkably ineffective in clarifying these questions and was no more able to achieve this from its one-sided stance than science was. If real progress is possible, we can expect it only from collaboration between them. Let this be my excuse if I as a layman touch upon these questions, for they are intimately linked with the goal of my talk.

Let us look to the famous thinker quoted above, namely Immanuel Kant, for advice on the question whether matter is continuous or atom-istically constituted, which he discussed in his antinomies. Of all the questions assembled there he explains that both the case for and against can be proved by strict logic. It is strictly provable that the divisibility of matter can have no limit and yet infinite divisibility contradicts the laws of logic. Similarly Kant shows that a beginning and end of time, or a boundary at which space stops are just as unthinkable as absolutely in-finite extension or duration.

This is by no means the only occasion when philosophical thought becomes enmeshed in contradictions, rather we meet it at every step. The most ordinary things are to philosophy a source of insoluble puzzles. In order to explain our perceptions it constructs the concept of matter and then finds matter quite useless either for itself having or for causing per-ceptions in a mind. With infinite ingenuity it constructs a concept of space or time and then finds it absolutely impossible that there be objects in this space or that processes occur during this time. It finds insuperable difficulties in the relation of cause and effect, body and soul, in the pos-sibility of consciousness, in short in each and every thing. Indeed, in the end philosophy finds it quite inexplicable that anything exists at all, that anything can ever arise and change, and in this it sees a self-contradiction.

To call this logic seems to me as if somebody for the purpose of a moun-tain hike were to put on a garment with so many long folds that his feet become constantly entangled in them and he would fall as soon as he took his first steps in the plains. The source of this kind of logic lies in excessive confidence in the so-called laws of thought. It is indeed true that we could have no experience if certain forms of linking perceptions, that is forms of thought, were not innate to us. If we wish to call these forms laws of thought, they are of course a priori in the sense that they are in our minds, or if we prefer in our heads, prior to any experience. However nothing seems less founded than an inference from the a priori in this sense to absolute certainty and infallibility. These laws of thought have evolved according to the same laws of evolution as the optical apparatus of the eye, the acoustic machinery of the ear and the pumping device of the heart. In the course of mankind's development everything inappropriate was shed and thus arose the unity and perfection that can give the illusion of infallibility. So too the perfection of the eye, the ear and the arrangement of the heart calls forth our admiration yet we could not assert these organs to be absolutely perfect. Just as little must the laws of thought be taken as absolutely infallible. Indeed it is precisely they that have evolved for the sake of grasping what is necessary for life and practically useful. With this the results of experimental research show much more affinity than does the testing of our thinking equipment. It is thus not to be wondered at that the forms of thought that have become habitual are not quite adapted to the abstract problems of philosophy which are so far removed from what is applicable in practice, nor have yet become so from Thales till now. That is why the simplest things are to the philosopher the most puzzling and why he finds contradictions everywhere. These, however, are nothing but inappropriate and mistaken mental re-shapings of what is given, which itself cannot contain any contradictions. Therefore as soon as contradictions seem uneliminable we must immediately seek to test, extend and alter what we call our laws of thought which actually are nothing but inherited and acquired ideas, confirmed throughout the ages, for denoting practical requirements. Just as the inherited inventions such as the roller, cart, plough, have long since been joined by countless artificial ones that have been created with full consciousness, so here too we must rearrange inherited ideas artificially and consciously into a better order. Our task cannot be to summon data to the judgment-throne of our laws of thought but rather to adapt our thoughts, ideas and concepts to what is given. Since we cannot clearly express such complicated conditions except by words, whether written, spoken or thought in silence, we can say that we must collocate words in such a way as everywhere to lend the most fitting expression to the given, so that the connections we create between words are everywhere as adequate as possible to the connection in reality. As soon as we put the problem thus, its most appropriate solution can still present the greatest difficulties, but at least one knows the goal being aimed at and will not stumble over obstacles of one's own making.

Many inappropriate features in the habits and behaviour of living beings are provoked by the fact that a mode of action that is appropriate in most cases becomes so habitual and second nature that it can no longer be relinquished if somewhere it ceases to be appropriate. I express this by saying that adaptation overshoots the mark. This happens especially often with mental habits and becomes a source of apparent contradictions between the laws of thought and the world, and between those laws themselves.

Thus the lawlike character of the processes of nature is a basic prerequisite for all cognition; hence the habit of always asking for the cause becomes an irresistible compulsion and we even ask for the cause of everything's having a cause. Indeed people racked their brains over the question whether cause and effect represent a necessary link or merely an adventitious sequence, whereas one can sensibly ask only whether a specific phenomenon is always linked with a definite group of others, being their necessary consequence, or whether this group may at times be absent.

Similarly something is called useful or valuable if it furthers the living conditions of the individual or of mankind, but we overshoot the mark if we ask for the value of life itself, when for example it seems to us pointless because it has no purpose outside itself. Just so when we take the simplest concepts out of which everything is built, and try in vain to build these in turn out of simpler ones or try in turn to explain the simplest fundamental laws.

We must not aspire to derive nature from our concepts, but must adapt the latter to the former. We must not think that everything can be arranged according to our categories or that there is such a thing as a most perfect arrangement: it will only ever be a variable one, merely adapted to current needs. Even the splitting of physics into theoretical and experimental is only a consequence of the two-fold division of methods currently being used, and it will not remain so forever.

My present theory is totally different from the view that certain questions fall outside the boundaries of human cognition. For according to that latter theory this is a defect or imperfection of man's cognitive capacity, whereas I regard the existence of these questions and problems themselves as an illusion. On superficial reflection it may of course be surprising that after recognition of the illusion the drive towards answering these questions does not cease. The mental habit is much too powerful to loosen its hold on us.

It is exactly as with ordinary sense illusions, which continue to exist even after their cause has been recognized. Hence the feeling of insecurity, the lack of satisfaction that grips the scientist when he philosophizes. Only very slowly and gradually will all these illusions recede and I regard it as a central task of philosophy to give a clear account of the inappropriateness of this overshooting the mark on the part of our thinking habits; and further, in choosing and linking concepts and words, to aim only at the most appropriate expression of the given, irrespective of our inherited habits. Then, gradually, these tangles and contradictions must disappear. What is brick and what mortar in the intellectual edifice must be made to stand out clearly and we should soon be freed from the oppressive feeling that the simplest is the most inexplicable and the most trivial the most puzzling. Unjustified modes of thought can in time recede, witness the fact that any educated persons now understand the theory of the antipodes and many that of non-Euclidean geometry. If therefore philosophy were to succeed in creating a system such that in all cases mentioned it stood out clearly when a question is not justified so that the drive towards asking it would gradually die away, we should at one stroke have resolved the most obscure riddles and philosophy would become worthy of the name of queen of the sciences.

Our innate laws of thought are indeed the pre-requisite for complex experience, but they were not so for the simplest living beings. There they developed slowly, but simple experiences were enough to generate them. They were then bequeathed to more highly organised beings. This explains why they contain synthetic judgments that were acquired by our ancestors but are for us innate and therefore a priori, from which it follows that these laws are powerfully compelling but not that they are infallible.

When I say that judgments like 'everything must be either red or not red' are derived from experience, I do not mean that every single person will test this uninformative proposition against experience, but that he gains the experience that his parents call every object either red or not red and he imitates them.

It may well seem as though we had gone too deeply into philosophical questions, but I think that the insight thus won could not have been reached by shorter and simpler paths, namely an unprejudiced judgment as to how to take the questions concerning the atomistic constitution of matter. We shall not now appeal to the law of thought that there could be no limit to the divisibility of matter, a law that is worth no more than if a naive person were to say that wherever he went on Earth the vertical seemed parallel to him so that there could be no antipodes.

We shall rather, on the one hand, start only from what is given, while on the other in forming and linking our concepts we shall heed only the aim of obtaining as adequate an expression as we can of what is given.

As to the first point, the most varied facts of heat theory, chemistry, crystallography indicate that in apparently continuous bodies space is by no means indiscriminately and uniformly filled with matter, but that it is occupied by great multitudes of individual objects, namely molecules and atoms which are extremely small though not infinitely so in the mathematical sense. Their size is calculable by many very separate methods all yielding the same result.

The fertility of this idea has been confirmed again quite recently. All phenomena observed in cathode and Becquerel rays and so on point to the fact that we are dealing here with tiny projected particles, namely electrons. After a fierce struggle this view was completely victorious over its initial opponent, the wave theory of these phenomena. Not only was the particle theory much better at explaining facts already known but it offered hints towards new experiments and allowed prediction of facts so far unknown and thereby developed into an atomistic theory of the whole theory of electricity. If it continues to develop with the same success as in the last few years, if phenomena like the transformation of radium vapour into helium, observed by Ramsay, do not remain isolated cases, then this theory promises to lead to as yet undreamt-of disclosures about the nature and composition of atoms. For the calculation shows that electrons are much smaller still than ponderable atoms, so that the hypothesis of atoms being built up out of many elements and various interesting views as to the mode of their assembly are now on everybody's lips. The word atom must not mislead us here, it has been taken over from ancient times, no physicist ascribes indivisibility to atoms today.

However, it is not all these facts and the consequences drawn from them that I wish to put forward here, for they cannot resolve the question as to the limited or infinite divisibility of matter. If we imagine what chemistry calls atoms as consisting of electrons, what would in the end prevent us from regarding these latter as extended bodies continuously filled with matter?

Rather, we will diligently abide by the philosophical principles developed above and therefore examine the formation of concepts itself in as unprejudiced a way as possible and seek to fashion it consistently and as appropriately as possible.

Here we find that we cannot define infinity in any other way than as the limit of ever growing finite magnitudes, at least nobody so far has been able to establish an intelligible concept of infinity in any other way. If therefore we wish to form a verbal image of the continuum, we must necessarily begin by imagining a very large finite number of particles endowed with certain properties and then examine the way in which this aggregate behaves. Certain of its properties may approach a definite limit when the number of the particles is made ever larger and their size ever smaller. We can then assert of these properties that they belong to the continuum and this in my view is the only non-contradictory definition of a continuum endowed with certain properties.

The question whether matter is atomistically constituted or continuous therefore reduces to the question: Which represents the observed properties of matter most accurately, the properties on the assumption of an extremely large finite number of particles, or the limit of the properties if the number grows infinitely large? Of course this does not answer the old philosophic question, but we are cured of the urge to want to decide it along a path that is devoid of sense and hope. The mental process, that we must start by examining the properties of an essentially finite aggregate and then let the number of items under it grow enormously, this process remains the same in both cases; it is merely an abbreviated expression of the same mental process expressed by algebraic signs if, as is often done, one starts from the differential equation itself in framing a theory of mathematical physics.

The several members of the aggregate that we choose as a picture for the material body can certainly not be thought of as being always absolutely at rest, for if so there could be no motion of any kind; nor even as being relatively at rest within one and the same body, because in that case we cannot account for fluids. Moreover no attempt has yet been made to conceive them otherwise than subject to the laws of mechanics. Let us therefore choose for the explanation of nature the aggregate of an extraordinarily large number of very small and constantly moving elementary individuals subject to the laws of mechanics. To this there has been an objection which we can suitably adopt as a starting point for the considerations that are to be the final goal of this talk. The fundamental equations of mechanics do not in the least change their form if we merely change the algebraic sign of the time variable. All purely mechanical processes can therefore occur equally well in the sense of increasing and decreasing time. But we notice even in ordinary life that future and past do not correspond at all so perfectly as the directions right and left, that on the contrary they are clearly distinguishable.

This is made more precise by the so-called second law of the mechanical theory of heat. It states that for any arbitrary system of bodies left to itself without influence from other bodies, we can always indicate in which sense any change of state will occur. For we can state a certain function of state of all bodies, namely entropy, which has the characteristic that any change of state can proceed only in a way that produces an increase in that function so that it can only grow larger with time. This law was of course obtained only through abstraction, like Galileo's principle; for it is impossible strictly to isolate a system from influence by any others. However, since this law combined with other laws has hitherto always given the right results, we consider it as correct, exactly as in the case of Galileo's principle.

It follows from this proposition that every closed system of bodies must finally approach a definite end state for which the entropy is a maximum. People have been amazed to find as an ultimate consequence of this proposition that the whole world must be hurrying towards an end state in which all occurrences will cease, but this result is obvious if one regards the world as finite and subject to the second law. If one takes the world to be infinite, the above-mentioned intellectual difficulties recur unless we imagine infinity as a mere limit.

Since in the differential equations of mechanics themselves there is absolutely nothing analogous to the second law of thermodynamics the latter can be mechanically represented only by means of assumptions regarding initial conditions. In order to find conditions that will serve this purpose, we must remember that for explaining apparently continuous bodies we must presuppose a very large number of all kinds of atoms or other general mechanical individuals to be in the most varied initial positions. In order to treat this assumption mathematically a special science was invented, whose task is not to find the motions of one individual system but the properties of a complex of very many mechanical systems starting from the most varied initial conditions. The merit of having systematized this system, described it in a sizable book and given it a characteristic name belongs to one of the greatest of American scientists, perhaps the greatest as regards pure abstract thought and theoretical research, namely Willard Gibbs, until his recent death professor at Yale College. He called this science statistical mechanics. It falls into two parts. The first investigates the conditions under which the externally noticeable properties of a complex of a very large number of mechanical individuals do not change, in spite of vigorous motions of these individuals; this part I shall call statistical statics. The second part calculates gradual changes in these externally visible properties if those conditions are not fulfilled, let it be called statistical dynamics. The wide perspectives opening up if we think of applying this science to the statistics of living beings, human society, sociology and so on, instead of only to mechanical bodies, can here only be hinted at in a few words.

To develop this science in detail would be possible only with the help of mathematical formulae in a series of lectures. Nor, quite apart from mathematical difficulties, is it free from difficulties of principle, for it is based on the calculus of probability. This latter is of course just as exact as any other branch of mathematics once the concept of equal probability is given. However, this being the fundamental concept, it cannot in turn be derived but must be regarded as given. It is as with the formulae of the method of least squares, which likewise cannot be set up unobjectionably unless we make certain assumptions as to equal probability. It is these difficulties of principle that explain why even the simplest result of statistical statics, namely the proof of Maxwell's law of velocities for gas molecules, continues to be disputed.

The theorems of statistical mechanics strictly follow from the assumptions made and will always remain true, like any well-founded mathematical theorems. However, their application to nature is the very prototype of a physical hypothesis. For if we start from the simplest basic assumption as to equal probability, we find for the behaviour of aggregates of large numbers of individuals laws that are quite analogous to those that experience shows to hold for the behaviour of the material world. Visible translational or rotational motion must increasingly transform into invisible motion of the smallest particles, that is into thermal motion, or as Helmholtz puts it characteristically: ordered motions transform more and more into disordered motion; the mixture of the various substances and temperatures, of places of greater or lesser molecular agitation, must always become more uniform. That the mixture was not complete from the start, but rather that the world began from a very unlikely initial state, this much can be counted amongst the fundamental hypotheses of the whole theory and we can say that the reason for it is as little known as that for why the world is as it is and not otherwise. However, one can adopt another attitude. States of great disentanglement, that is great temperature differences, are not absolutely impossible according to the theory, but merely highly improbable, although in an almost inconceivable degree. If only, therefore we imagine the world as large enough, then according to the calculus of probability there will supervene now here now there regions of the dimensions of the system of the fixed stars that have quite an improbable distribution of states. Both during their formation and during their dissolution the temporal course will be uni-directional; if there are intelligent beings in such a location they must gain the same impression of time as we do, although the temporal course of the universe as a whole is not uni-directional. This theory does indeed boldly transcend experience but it has precisely the feature that such theories should have, namely to show the facts of experience in quite a new light and thus to spur one on to further reflection and research. In contrast with the first law of thermodynamics the second appears only as a probability theorem, as Gibbs had stated already in the 1870s.

I have not shirked philosophic issues here, trusting that philosophy and natural science, working in concert, will give each other fresh nurture; only so can we truly express ideas consistently. Schiller told the scientists and philosophers of his day: "Let strife divide you, it is too early for a pact". I do not dissent, but think the time for a pact has come.

REPLY TO A LECTURE ON
HAPPINESS GIVEN BY PROFESSOR OSTWALD*

Schopenhauer prefaces his critique of Kantian philosophy by an introduction in which he declares he must begin by expressing once and for all his great admiration for Kant, so that for brevity's sake he might afterwards confine himself to discussing what seems defective to him without constantly interrupting the argument by reaffirming his admiration and emphasizing the many excellent things that Kant discusses along with the seemingly incorrect material. I shall adopt the same procedure here, by first expressing my personal thanks to Prof. Ostwald for the great pleasure and intellectual inspiration that I have derived from his many-sided, profound and original writings and lectures, but then addressing myself without further ado exclusively to those parts with which I disagree.

Let me observe further that such a controversy can never have as its purpose to exhaust the topic much less to force a decision as to who is right and who wrong; as a rule neither is absolutely right or absolutely wrong. The purpose of controversy is rather to illuminate the topic from all sides and to inspire the debaters as well as the audience to further reflection. That is why after a possible reply I renounce any rejoinder from the outset.

For a long time now, and certainly long before Rankine introduced the term energy in its present meaning into science, a forceful exercise of the will used to be called energy. Let us call it mental energy in contrast with the physical energy of Rankine. We thus have the same word for two objects, but it remains questionable whether the meaning is the same in the two cases; I rather doubt it. In natural science energy is a magnitude that can be measured exactly, plays a role in various fields, but as soon as measured in suitable units always maintains itself in quantity so that if it vanishes in one place an equal amount always appears somewhere else.[1] Only if it had been proved that when mental energy is developed an equivalent amount of physical energy always actually disappears, that is if mental energy can be measured in such units that the amount developed

* *Populäre Schriften*, Essay 20. Delivered to the Vienna Philosophical Society in 1904.

was always exactly equal to the physical energy lost, should we be entitled to speak of mental energetics.

The proof of this proposition has however by no means been achieved, indeed everything points to it being impossible, and that because it is completely false. The perfect parallelism between mental phenomena and physical brain processes makes it probable that all energy is constantly maintained in its physical form within the brain mass, while mental processes are merely parallel epiphenomena without energy, indeed perhaps merely a second mental picture of the same phenomena viewed from another angle, which can thus certainly not contain any new sort of energy in the physical sense.

If we really assigned to mental phenomena a new form of physical energy, namely mental energy in Ostwald's sense, and assumed that the two kinds could be transformed into each other according to the conservation principle, we should be led back to the perennial doctrine of a special psyche, existing along with the body and able to provoke motion in parts of the brain mass or other parts of the body, like a magnet acting on soft iron; this is a view that most physiologists thinking clearly along scientific lines and by now most of the most clear-minded philosophers too would regard as improbable.

Be that as it may, even if we were to assume such mutual action between body and mind, it remains certain that what is called energy of will is something quite different from what is called energy in natural science. Imagine a very energetic man, he begins by walking up and down in his room and makes decisions, then he conveys them to the members of his family, his friends and his subordinates in clear and decisive words, managing to ensure that they all carry out what he was aiming at. All these processes no doubt require a certain amount of physical energy, since they are accompanied by physical processes of the brain mass and the limbs. Let us now compare this with the case of a neurasthenic who madly runs about his room, storming and cursing, scolding and shouting at those round him, and all this merely because he is in doubt whether the weather will remain good and he cannot decide whether he is to go for a walk or stay at home. Are not all the indications that his activity consumes as much if not more physical energy than that of a strong-willed man, and yet it is the latter who produces the highest mental energy and the former none.

To defend Ostwald's view one might however argue that the energy spent on moving the limbs during walking in the room, on moving the larynx, lung, tongue and so on during speech and on the brain functions involved in producing thought were indeed the same in the two cases; but apart from this there was a numerically definite amount of physical energy that became transformed into a quite new form, namely mental energy in Ostwald's sense, which is perfectly interchangeable with physical energy. To refute this would of course be just as difficult as to prove it. At all events it is premature to infer from the fact that it has become common usage to call both by the same name, that mental energy corresponds to an equivalent amount of physical energy and is thus subject to the conservation principle, which did not suddenly appear from thin air as regards physical energy, but was not recognized as a law of nature until the most extensive and laborious experiments had shown it to be correct.

Moreover, the mental form of energy could only very temporarily reside apart from physical processes of the body and would always quickly revert back into physical form; for otherwise a large amount of energy would in time have to exist outside physical processes of bodies and at death would suddenly have to reappear as purely physical energy, as heat or perceptible in some other way, unless, indeed, one were to assume that the psyche takes its energy into the beyond which is inhabited not only by spirits but also by beings that suffer changes subject to Robert Mayer's law of the conservation of energy.

If, however, physical energy and what I called mental energy are two totally different things called by the same name because of a rather superficial similarity, I think it is mistaken, because productive of false ideas and leading to error, when people speak without distinction of an energetic theory of mechanics, chemistry, mental phenomena, happiness and so on.

In all his writings, Ostwald expresses great appreciation of Mach, and indeed rightly so; my respect for Mach is none the smaller, even if I do not share his views on everything. However, as regards Ostwald's energetics, I think it rests merely on a misunderstanding of Mach's ideas. Mach pointed out that we are given only the law-like course of our impressions and ideas, whereas all physical magnitudes, atoms, molecules, forces, energies and so on are mere concepts for the economical representation and illustration of these law-like relations of our impressions and ideas. These last are thus the only thing that exists in the first instance, physical concepts being merely mental additions of our own. Ostwald understood only one half of this proposition, namely that atoms did not exist; at once he asked: what then does exist? To this his answer was that it was energy that existed. In my view this answer is quite opposed to Mach's outlook, for which energy as much as matter must be regarded as a symbolic expression of certain relations between perceptions and of certain equations amongst the given phenomena.

As regards the concept of happiness, I derive it from Darwin's theory. Whether in the course of aeons the first protoplasm developed 'by chance' in the damp mud of the vast waters on the Earth, whether egg cells, spores or some other germs in the form of dust or embedded in meteorites once reached Earth from outer space, is here a matter of indifference. More highly developed organisms will hardly have fallen from the skies. To begin with there were thus only very simple individuals, simple cells or particles of protoplasm. Constant motion, the so-called Brownian molecular motion, happens with all small particles as is well-known; growth by absorption of similar constituents and subsequent multiplication by division is likewise explicable by purely mechanical means. It is equally understandable that these rapid motions were influenced and modified by the surroundings. Particles in which the change occurred in such a way that on average (by preference) they moved to regions where there were better materials to absorb (food), were better able to grow and propagate so as soon to overrun all the others.

In this simple process that is readily understood mechanically we have heredity, natural selection, sense perception, reason, will, pleasure and pain all together in a nutshell. It merely requires a change of quantitative degree with constant application of the same principle, to proceed via the whole world of plants and animals to mankind with all its thinking, feeling, willing and acting, its pleasure and pain, its artistic creation and scientific research, its nobility and vices.

Cells that had come together into rather large collections in which there occurred division of labour and hived off, by division, cells with similar tendencies, had greater opportunities in the struggle for existence, especially if certain cells when subjected to harmful influences did not rest until the working cells had removed such inroads as much as possible (pain). The action of these cells was particularly effective if, so long as the harmful influences were not completely eliminated, it continued and left behind a tension that diminished only very slowly, thus stressing the memory cells and inciting the motor cells to even more vigorous and circumspect collaboration when similar harmful circumstances recurred. This state is called lasting displeasure, a feeling of unhappiness. The opposite, complete freedom from such vexatious after-effects and a warning to the memory cells that the motor cells are to act in the same way again if similar circumstances supervene, is called lasting pleasure, a feeling of happiness.

This does not of course exhaust all the gradations of these feelings in highly organized beings. Not even a beginning has been made for a physiology of happiness; but the point of view has been fixed from which the relevant phenomena have to be regarded if one is to explain them scientifically, rather than merely turn beautiful, uplifting, poetic and inspiring phrases about them.

In this we naturally consider only one side of affective phenomena, namely that which can be grasped by natural science. We see why the processes of an organism built very similarly to our own touch us much more directly and appear in quite a different light than those of a totally different one, so that we should never say of a machine made by men from rods and wheels that it was happy or unhappy, even if it were as complex and centrally organized as our own organism and similarly incited by external influences to act in appropriate ways; this is indeed an idea that we find much more difficult to imagine ourselves into than would be thought by those who adhere to the hypothesis that there are mental phenomena existing apart from brain processes.

This moreover makes it perfectly clear that for every individual the only immediately given consists of his own mental phenomena (but not the brain processes which, though identical with them, are not recognized because they are mechanical), while atoms, forces, and energy forms are mental concepts constructed much later in order to represent the law-like features of perceptions.

However, I could never understand how it could be said that we felt immediately that our sensations were not merely a consideration of purely physical processes from another angle, but must be something quite different from them and newly added to them. In this way people before the time of Columbus thought they felt immediately that the antipodes were impossible, and before Copernicus that the Earth was not revolving.

Ostwald expresses the magnitude of happiness by means of the algebraic formula $E^2 - W^2 = (E + W)(E - W)$, where E denotes the energy spent intentionally and successfully, and W that spent with dislike.* On this I must comment that a genuine mathematician puts definite powers into a formula only if it has been found by exact measurements that just this and no other power is required to obtain agreement with experience. Has Ostwald proved that $E^4 - W^4$, $E^n - W^n$ or many similar formulae agree less well with experience?

That besides the difference $E - W$ the sum $E + W$ also contributes to happiness is the conviction of an enterprising Western European. A Buddhist, whose ideal is the mortification of the will would perhaps write $(E - W)/(E + W)$. Besides there are mathematical formulae where the operations are meant only symbolically, but then the applicability of any law of calculation must be proved afresh. If the formula is meant only symbolically, it is no longer evident that the two expressions $(E + W) \times (E - W)$ and $E^2 - W^2$ are actually equal, that the rule for multiplication for algebraic magnitudes is applicable to these symbolic expressions too, rather this must first be specially established.

On the other hand Ostwald's formula does not mention certain quantities on which happiness evidently depends, for example the immediately preceding circumstances of happiness. Regard for these led my brother Albert who died while still at high school to the following definition of happiness: a person's happiness is equal to the degree of agreeableness of what he is thinking at the moment, minus what he regards as the average degree of agreeableness of what he would be thinking were he not thinking what he is thinking (in the style of Behrisch in Goethe's *Truth and Fiction*, two pages before the start of book 8).**

This essay was written by request of the editor of the "Umschau", who wished to publish in his journal the fleeting words that I had spoken in reply to Professor Ostwald's lecture on happiness given at Vienna in 1904. I believe I was not the only one at the time who felt that Ostwald had half allowed himself a joke, and I replied in the same spirit. Surely, too, it must appear as a joke that I published my reply long before Ostwald's lecture was printed. For the sake of completeness I have included it here, but now that Ostwald's lecture has appeared in print I can only regret my facetious tone. For when a scientist with Ostwald's reputation and influence deals such a blow to the exact method which has evolved over many centuries and has proved itself as the only method leading to the goal, it is a deadly serious matter. I therefore beg the reader to let me add a few complementary remarks to the above essay.

* *Editor's note*: E from *Erfolg* (success), W from *Widerwillen* (dislike).
** *Editor's note*: Behrisch told the young Goethe that true experience came with experiencing how an experienced man experienced experience when he experienced it.

Some years ago I had already occasion energetically to oppose Ostwald's energetics on purely physical territory. If now I do the same thing again it is certainly not for personal reasons; indeed I believe I am happy enough to be able to count myself amongst his best friends and I admire his work in physical chemistry, nor am I in principle opposed to attempts at building a theory that puts the concept of energy at the head, merely to Ostwald's way of doing it.

If therefore I now put forward, so far as I am able, a W against Ostwald's energetics' E, this is done merely because I cannot banish the thought that a return to the inexact method of Ostwald's essay on happiness, which people thought had finally been overcome, would mean for science a relapse by several centuries.

After Ostwald's present explanations there can be no doubt that he speaks of energy in the ordinary physical sense of the word. The total energy C, produced in the organism by oxidation of matter taken in as food and transformed in part directly into heat and partly into mechanical energy, Ostwald begins by dividing into two parts, namely D, which is used for unconscious physiological functions (maintenance of body heat, circulation of the blood, breathing, digestion and so on), and $E + W$, whose transformation involves conscious actions. The first he leaves completely out of account, adducing only the second in his considerations about happiness.

Right at the beginning of the discussion of this magnitude $E + W$ it seems that an unconscious memory of the other psychological sense of the word energy discussed above already plays a nasty trick on him. Because what we call psychological energy is most intimately connected with efforts of the will, he thinks it probable that $E + W$ is proportional to the strength of will. To prove this hypothesis he merely says that a tired brain is incapable of efforts of will and that unaccustomed efforts tire us. He further admits that a personal factor is involved as well, so that with different people the same act of will corresponds to very different expenditures of energy derived from combustion of food. However, I regard Ostwald's hypothesis that even for the same person there is so much as a shadow of proportionality between physical energy consumed $E + W$ and strength of will as absolutely misguided.

It seems to me that the will everywhere merely has the character of a triggering agent that initiates energy transformations but its intensity seems to be as little proportional to the amount transformed as the intensity of a spark exploding a powder keg is to the amount transformed in the explosion. I can walk up and down in my room, go for a walk, climb a mountain, all of which is done consciously with efforts of the will, but my intensity of will can be very slight. The most insignificant circumstances would induce me to refrain from these actions, although they involve a large transformation of energy. I apply very little energy in the psychological sense, but much in the physical.

On the other hand I might in the highest degree desire the solution of a mathematical problem important to me or the attainment of an honorific post or release from bodily pain and so on, but my reflection is accompanied by a very small amount of purely physical energy. Success in solving the problem makes me very happy, refraining from climbing the mountain would not make me unhappy. From these considerations it follows that it is not the quantity $E + W$ of physical energy transformed that is decisive for the intensity with which one wills something.

Surely, however, the essence of energetics cannot lie in attaching the word energy to everything, whether the term, which has a definite significance in physics, fits the occasion or not. Evidently the quantity of energy transformed during acts of the will has nothing to do with the magnitude of happiness: that is influenced only by the actual intensity of will which is something totally different.

The case is very similar with the way Ostwald splits off from the total energy $E + W$ the part W spent against our will. If something happens against our will it is unpleasant to us, it contributes not to our happiness but to unhappiness. To see this we need no energetics, but here too I regard the quantity of physical energy spent against our will as the most inappropriate measure possible. The unpleasantness is proportional to anything rather than to the energy consumed against our will measured in physical units. With very little outlay of energy we can make terrible mistakes that are fatal for our whole life and with a very large amount of energy attract very insignificant inconveniences.

Ostwald himself says in one place that what matters is not the real resistance but only our mental feeling of a resistance, and the latter in my view has nothing to do with energy except that it is connected with physico-chemical processes in the brain and in the external world which are not possible without transformation of energy. However, there is not the slightest trace of proportionality of feeling and energy expended or of the one being measurable by the other.

It seems once again that W is being called energy only because it is a principle of energeticists to call everything energy whether it is proportional to mechanical energy or not.

Incidentally I think that there are other reasons too why this splitting of the total energy consciously used, $E + W$, into the two parts E and W is not as simple as Ostwald imagines.

This makes us think automatically of a weight that sometimes falls (as it were according to its will) and sometimes is raised (against its will). However we must immediately reject this analogy, since the weight performs positive and negative work respectively without containing a source of work, whereas a man in oxidizing food contains such a source which he transforms with or against his will into work of the same sign. Therefore an energy turnover set off directly by my will must always proceed according to my will, only with later, secondary effects can it become questionable whether they answer my will or not.

When I integrate a differential equation the movements of the pencil always follow the impulses of my will, only the final result may be different from what is desired. Thus it is often not the energetic activities as such that determine our happiness or otherwise but the later, secondary consequences that are independent of the will. Indeed the activities involving energy cannot as such be divided into those that correspond to our will and those opposed to it, only the later consequences can, and they no longer involve our own expenditure of energy. It is not an energy outlay against our will that makes us unhappy, but only the conviction (or perhaps sometimes the fear) that our activities might subsequently fail to lead to the consequences we desire. If from fear of punishment or the threat of some other disaster we are compelled to expend energy against our will, the feeling of unhappiness does not grow with the energy expended; the feeling of unhappiness is greater still if we can do nothing to ward off the disaster.

That too is the reason why things that do not depend on our will at all contribute not only to our momentary well-being but to our actual feeling of unhappiness; for example bad digestion or a liver complaint contribute to unhappiness, a glass of good wine, and according to Ostwald progressive paralysis too, to happiness. Of course Ostwald says that this is because in the former case W seems to us increased and in the latter diminished. The unprejudiced observer will hardly be able to deny that it is the other way round. It is not because W seems greater to him that the sufferer from liver disease is unhappy, but it is because he feels unhappy through purely physiological agencies that have not the least notion of being proportional to an energy outlay either with or against the will, that W seems to him magnified and everything rather dreary. If now he succeeds by means of taking pills in eliminating the disease, he may have spent little energy in doing so and yet he will have greatly improved his feeling of happiness. It would thus be more sensible not to split energy into portions expended according to and against the will, but to multiply every amount of energy by the degree of volition, to be assumed positive in the first case and negative in the second, and then to put the sum of these products in place of Ostwald's $E-W$; however this too would not do, since what is willed is not the energy outlay, but its consequences. The explanation why with bad digestion W seems so large and under inebriation or paralysis so small is something Ostwald still owes us.

I could make a great many other more detailed remarks. For example, the formula should express displacements of the zero reference level from which $E-W$ is measured, displacements that depend on subjective feelings and personal factors, for the purpose of a formula surely is to express the unknown by the known, not by further unknown features. Likewise, a formula that claims utility should contain the universally known aftereffects of prior happiness or unhappiness on our present feeling of happiness, as on all our other feelings, something not contained in Ostwald's formula but only in the attendant notes. Surely, the sudden discovery of an object we thought lost makes us happy, just as after a period in a totally dark room a normally lit one will seem blindingly bright. What is the use of a formula if a circumstance that is so important for the momentary feeling of happiness is not expressed in it but has to be added as verbal notice afterwards!

However, I am afraid I should bore you were I to go into further detail. Let me therefore briefly sum up. Analysing the matter without prejudice, the whole content of Ostwald's formula seems simply to be this: we feel the happier the more (E) occurs according to our will and the less (W)

against.[2] Ostwald of course adjoins the factor $E + W$; that is, he asserts that the more energetic of us feel happier when happy and unhappier when unhappy than the less energetic. This would hardly be an epoch-making discovery. Moreover, it would still have to be proved; Buddhist saints might assert the opposite. Besides we must remember that what we are dealing with here is not moral energy, but the physico-chemical variety, proportional to the heat of combustion of nutriments, so that this factor is of value mainly to the strong men of the fairground and for people doing heavy physical work.

It seems to me that none of Ostwald's considerations grow organically from his formula, that they fail to constitute an analysis of this formula in the sense of analytic geometry or mechanics; rather, they merely stand in some loose connection with it. I would say, the name of energy is taken in vain throughout this whole dissertation. It is as though somebody were to say that the beauty of music was measured by $(E - W)(E + W)$, where E is the acoustic energy expended in agreement with good taste and W that expended against it, while the factor $E - W$ is to express that music is the more beautiful the more it corresponds with good taste, and the factor $E + W$ that strong music as such generally has a stronger effect than music that is too weak. Of course deafening music would again run counter to good taste, here W would be again very large and $E - W$ small or even negative.

Why is it that an apparently harmless article like Ostwald's seems to me so dangerous for science? Because it marks a relapse into delight in the merely formal, into the pernicious method of so-called philosophers, the method of constructing theoretical edifices from mere words and phrases and placing emphasis only on their pretty and formal interlacing; and this used to be called a logical foundation, without regard whether this lacework corresponded with reality or was adequately grounded in fact. It is a relapse into the method of submitting to the dominion of precon-ceived opinions, of bending everything to the same principles of classi-fication, of trying to force everything artificially into the same system. It is a refusal to see true mathematics for mere formulae, true logic for mere seemingly correctly constructed syllogisms, true philosophy for mere bric-a-brac in philosophic guise, the wood for mere trees. This method, un-fortunately, will always please the masses more than does the method of natural science, which leaves less room to fantasy.

NOTES

[1] Thermal energy can be transformed into electrical, chemical into thermal and so on.
[2] That is why I can hardly imagine that anyone might have obtained from this formula practical hints useful for life and contributing to this happiness. For the formula merely says 'be energetic and see to it that everything happens according to your will'; and this much, I think, everybody knows even without a mathematical formula.

ON A THESIS OF SCHOPENHAUER'S*

A writer once said that the most important thing in a literary work is to give it the right title: with a novel or play success was jeopardized if the title was badly chosen. If the same holds for philosophical talks, I am in dire straits today.

I wish to speak about Schopenhauer; in order to give my talk some local colour, I wanted to imitate Schopenhauer's style in my very title. His style distinguishes itself especially by the mode of expression associated in the past with fishwives but which today might be called 'parliamentary'.

In this spirit I had chosen for my talk the following title: 'Proof that Schopenhauer is a stupid, ignorant philosophaster, scribbling nonsense and dispensing hollow verbiage that fundamentally and forever rots people's brains'. These words are quoted verbatim from *The Fourfold Root*[1], except that there they relate to another philosopher. Schopenhauer's style was excused as due to his anger because he had been passed over when posts were filled. Question: is his anger or mine the holier, given the object of it?

Well, this title was denied me, and rightly so. For what was I to have offered in my talk, if in the title I had already blown all my powder, indeed one might well say dynamite? However I was too indolent to seek a second, thoroughly adequate title, and thus arose the title printed in the introduction, not quite in conformity with the matter.

I wish to speak not on a thesis of Schopenhauer's but about his whole system, though certainly not to furnish a complete critique but merely some sketchy thoughts on the subject.

What I have to say may well not be new, but to verify this I should have to go through the works of the various philosophers and there I am in some embarrassment, for I am not really quite sure what philosophy is. This happens with other sciences too, that we cannot give a strict definition of its concept, yet one knows the objects with which they deal. With phi-

* *Populäre Schriften*, Essay 22. Delivered to the Vienna Philosophical Society, 21 January 1905.

losophy, however, I am not even sure whether it differs from other sciences by its objects, for example if it were the investigation of mental phenomena, or whether it differs only by its method.

I will not go too deeply into this question, rather I understand by philosophers those who have hitherto commonly been called so, irrespective of definitions of philosophy.

In the works of these philosophers there is much that is apposite and correct, for example their remarks when they scold other philosophers, but what they themselves contribute mostly lacks these qualities. If therefore I now put forward various things against Schopenhauer, I am convinced that much of it is to be found already in other philosophers. I can only hope that what I said about the novel additions of other philosophers does not hold of what might be new in my remarks.

Everyone knows the perennial controversy between idealism and materialism. Idealism asserts that only the ego exists, the various ideas, and seeks to explain matter from them. Materialism starts from the existence of matter and seeks to explain sensation from it.

Schopenhauer tries to transcend these oppositions by saying that the existence of the whole world rests on subject and object, neither of which is anything at all on its own, but they exist only in their mutual relation: an object can exist only relatively to its subject and vice-versa.

What makes this even more complicated is his assumption that the subject can be its own object. In that case we are again left with a subject and no object. He clarifies this as follows: the cognizing subject cannot be object, but the subject as willing can be object, so that the subject is split into a cognitive and a volitional one. The volitional is object of the cognitive. Questions as to a more detailed explanation he stops short by saying that this is the cosmic knot that cannot be disentangled.

Schopenhauer then proceeds to what Kant calls forms of the intuition, time and space. I quote:

If time were the sole form of the intuition there would be no simultaneity and therefore nothing constant, no duration. For time is perceived only if it is filled and its progression by the change of what fills it. The constancy of an object is therefore recognized only by the contrast with the change of other objects that are contemporaneous with it; the idea of contemporaneity is impossible in pure time....

If on the other hand space were the only form of the intuition, there would be no change; for change or transformation is a succession of states and succession can occur only in time. Therefore time could equally be defined as the possibility of opposite determinations of the same thing. Thus we see that the two forms of empirical ideas, though sharing infinite divisibility and extension, are fundamentally different, since what is essential to one has no significance in the other, juxtaposition none in time and succession none in space.

I think you will admit that there is little actual content in this. It really says merely 'time is time and space is space'.

We have indeed made considerable progress in the modern theories of space and time as against this point of view of Schopenhauer. Above all, space has been construed by means of the concept of number alone, without any help from intuition; it was possible to ascertain which properties remain to space and which become changed if we drop this or that geometrical axiom, and thus what are the spatial experiences with which each of the axioms is particularly connected and that none of the axioms is really evident a priori.

Quite in general, Schopenhauer was not at all felicitous in what he called a priori. For example he says that it is a priori clear that space has three dimensions. Today scientists know that 'a priori' a space of more than three dimensions is conceivable and even a non-Euclidean one. Of course the question is not whether the space of experience is Euclidean or not, but what is evident a priori and what merely a matter of experience.

Similarly Schopenhauer infers from the principle of sufficient reason that the law of conservation of matter is clear a priori. Landolt has conducted experiments on precisely this law, and his findings seemed at first to contradict it. Today it is indeed more likely that they will not be able to impair the law of conservation of matter, but what is important here is not the results of his experiments but merely whether experiments are as such able to refute laws, or whether logic can prescribe the paths that the pointer on Landolt's scales must take.

For a second time doubts about this law have arisen in connection with the behaviour of radium. I am convinced that these experiments too will confirm the law, but that proves the law to be other than a priori: were it not to hold, we could retort nothing from a logical point of view.

Schopenhauer relies on the fact that, were the law to be invalid according to experience, we could never attain the concept of matter, since matter is for all of us that which persists and only so can we attain the concept. However, it does not follow from this that there can be no exception. If Landolt's experiments were to demonstrate the opposite, we should have to change our idea of matter and regard it as generally persisting, but variable in certain isolated exceptional cases.

Let me now proceed in particular to the role played by the will in Schopenhauer's philosophy. He says that "when a stone falls to the ground this is as much an act of the will as when I myself will something; but since I am within myself I know that it is an act of the will. If I could look into the innermost essence of the stone, I should see that it too has a will". This is quite an ingenious remark, but if Schopenhauer is now firmly convinced that by using the same word will for forces of inorganic nature and for certain psychological processes that we experience in ourselves he has made colossal strides in our knowledge of nature, he really yields to a rather naive illusion. We shall do better if we reserve the word 'will' for the conscious drive towards action in man and the higher animals and refrain from applying it to plants and stones, in order to have a characteristic word for each of these phenomena without fear that we might make less sense of it than Schopenhauer with his way of looking at things.

In a way that is even more strange, Schopenhauer introduces the concept of 'freedom'; the will as subject, as thing in itself, is necessarily and unconditionally free, since causality has no purchase on things in themselves. It is completely free under different external circumstances to act quite differently. However the actions of the will, its manifestations, its objective realization under given circumstances are completely determined by these latter and thus completely unfree and from the freedom of the will as thing in itself we can explain the obscure feeling that our actions too are free.

For these actions too, however, he leaves open a back door: if the will aims at its own destruction it no longer depends on anything and a moment of freedom supervenes.

That Schopenhauer's considerations are ingeniously contrived and inspire to a lively exercise of wit and reflection without however containing anything of permanent truth, is shown in their application. He applies them to the different arts: these are supposed to be the liberation of the will from objectivity, its refinement from anything distinctive and special. Architecture is the art of solid bodies and works with them; in contrast he mentions hydraulic arts that work with liquids, the art of horticulture which works with men and animals. An art that works with gases he simply forgets. That would have to be the art of fanning air to one's fellows or an art that works with smells and thus has the task of inspiring the olfactory sense in an artistic manner; I am not quite sure which. Such an art does not exist, but from this it does not follow that it would be logically contradictory. However, the culinary arts would be bound to come first in importance, or better expressed, the artistic influence on the sense of taste.

Next Schopenhauer mentions a series of arts that imitate nature visibly: thus the plastic arts, though as to independence, tailoring and hairdressing are perhaps superior to it; thus likewise the painting of landscape, plant and animal still-life, and following on from there the painting of animals and human beings which along with the sculpture of animals together first represent moving objects though only in one phase of motion. Painting of humans is the art of portraiture or of imagined dramatic or historical scenes in which however only the purely human value of the representation is considered and not the historical.

The transition to the art of the writer is formed by symbolic plastic art and painting, since he acts only by means of thought symbols. The most subjective form of it is lyric poetry, next come all forms of poetic tales in prose and verse, and finally the most objective poetry, namely drama, which however enlists the help of painting, music, the dance and representational skill of the actor which last must be described as the most perfect plastic art, compared with that in stone or bronze.

Finally we come to music. According to Schopenhauer this is the direct representation of the will in so far as it is not object, whereas all other arts also represent will but only indirectly as an individual objective form of it. Since we cannot perceive will, but only will become object, we cannot analyse music intellectually. I would agree that music has something very special over all the other arts. However, in its details Schopenhauer's theory is untenable. Some of it is indeed rather funny. For example, if the ground bass is supposed to resemble the mineral kingdom, the lower intermediate voices the kingdom of plants, the higher ones the animal kingdom and the descant the realm of man. According to Schopenhauer music is a mirror of the world but not its pictorial representation, not even partially so as with the other arts, but it stands on a level of equality with the world as a whole, the world being one manifestation of the cosmic will and music another, achieved by different means and independent of the first. Hence too the paradoxical proposition that music could continue to exist if the world did not. Of course there would not then be violins, sound-transmitting air, excited ears, perceiving minds. However these are only the means of art, just as brushes, paintpots, palettes, canvas, the luminous aether, the eye and the psyche of the spectator are in the case of painting. In addition the painter needs the object that he copies. Not so music, which by its own means directly creates a picture of the cosmic will. Of course one might say the same thing of the art of pyrotechnics or of ornamental art that does not copy plants or of architecture that does not serve practical ends or indeed of the art of dancing.

As in the highest things, Schopenhauer's extravagance equally shows itself in the most trivial. For example he has a dreadful antipathy towards the male beard. It is something bad, and that for philosophical reasons. First, hairy surfaces remind us of the animal kingdom, and therefore the male must relinquish the cover on the lower half of his face. Secondly, the beard prolongs that part of the face that represents animal features and contains the organs for chewing. This part of the face should be limited. Thirdly, beards are supposed to be completely dead matter devoid of nerves and muscles and it offends taste to carry so much dead matter about with one.

In this way Schopenhauer seeks to give an aesthetic basis to his views. An explanation that seems less remote would have been that one of his opponents, perhaps one who had opposed his appointment to a professorship, had worn a long beard. We see how a philosopher who regards aesthetics only from a theoretical angle can go astray. The result, using Schopenhauer's mode of expression, is this: "Stupidity, simple-mindedness, foolishness, mental daubing, folly, eccentric nonsense, cranky obtuseness, imbecility that cries to high heaven". I hope that this load of dynamite is enough.

Now I come to the topic which could most readily have been meant by the word thesis mentioned in the title, namely ethics.

From his whole theory of the will, Schopenhauer deduces the consequence that life is a misfortune.

For nothing exists except will, but will must always will something, strive after something. As long as what it strives after is not attained, the will is dissatisfied, unhappy, when it has attained its goal the will ceases and happiness with it. Either one then strives after something new and so is again dissatisfied, or boredom supervenes, the worst state of all. Therefore one cannot help oneself, life is always unhappy. The only correct ethics consists in the will's denying itself and in one's preparing the transition to nothingness. That then is happiness.

This derives from an old Indian theory which makes the strange inference that being cannot be split into several things. For otherwise one part would have to be as another is not. But it is contradictory for something that is to lack being in any respect. Moreover what is cannot change, for then it would have to be now what it was not earlier, and again what is would be lacking being. The truly being must therefore be something one and unique for ever indivisible and immutable.

Next one notices that nothingness possesses all these properties. Nothingness is one, there are not several of them, nor does it change with time. Therefore it is really nothingness that is; everything that we take to have being, that is for ever split and ununited with itself involved in internal warfare, incomprehensibly vanishing again in the very moment of its birth, all this is really nothing. But because we ourselves are nothing we are unable to lift the veil of Maya and regard nothingness as something while taking what truly is for nothingness. This is Schopenhauer's view too. The dissolution into nothingness he tries to sweeten for us by representing it as transition into true being.

Further detail emerges if we go a little deeper into the theory of nothingness. We must distinguish first the privative nothingness which is nothing only relatively to certain things. For example I expect that a little box contains jewellery and remark with disappointment that there is nothing in the box, although it contains luminous aether, atmospheric air and perhaps even cotton wool. Secondly, there is negative nothingness, which is already more nothing than the first variety, something that is naughter than naught. For example, I consider a really empty room, free even from luminous aether. However this is still a mental thing, only relatively nothing and we can oppose to it the idea of absolute nothingness, a nothingness that is actually nothing at all, the nirvana or 'prachna paramita' of the Indians. Anyone who philosophizes in this manner must indeed feel flattered if one says that nothing can come of it, for it is precisely this that he regards as the real something.

However, let us leave aside these theoretical speculations as to whether the concept of nothingness is merely relative or whether something could be nothing merely relatively towards something else; let us rather ask for the more practical consequences, and then it is precisely the theory that ethics should lead us to strive after nothingness and renunciation that is misguided. If the Germanic tribes adopted this ethics they would become Hindus and the other peoples would overrun us.

However, men have been intelligent enough not to believe Schopenhauer. In my view it is utterly wrong to regard it as the task of ethics to deduce from metaphysical arguments whether life as a whole is a happy or an unhappy circumstance. For everybody this is a question of his own subjective feelings, his bodily health, his external conditions. No unhappy person is helped if we go on to give him metaphysical proof that life itself is a disaster. If, however, we are looking for means to heal or alleviate physical or moral weaknesses, we can at least actually help some unhappy people.

Ethics must therefore ask when may the individual insist on his will and when must he subordinate it to that of others, in order that the existence of family, tribe, or humanity as a whole and thereby of each individual is best promoted. This innate love of asking, however, overshoots the mark if we ask whether life as such should be promoted or inhibited. If some ethic were to cause the decline of a tribe adhering to it, that would refute it. In the last instance it is not logic, philosophy or metaphysics that decide whether something is true or false, but deeds. "In the beginning was the deed", as Faust says. What leads us to correct deeds is true.

That is why I do not regard technological achievements as unimportant by-products of natural science but as logical proofs. Had we not attained these practical achievements, we should not know how to infer. Only those inferences are correct that lead to practical success.

Of course once a method of inference has been tried and inherited through thousands of years, it seems to us a priori correct, so that we often can go on working with them for long stretches without practical tests, for example if we are confident that a calculation will yield the correct result; however, at some time it must have been tested by deeds, and from time to time it must be tried again.

Now it seems to me that just as Schopenhauer's ideas have shown themselves to be untenable, so too are those of all other philosophers in their central core, Kant included, although naturally I have not time enough to prove this here.

The question now arises: was the work of these great minds really in vain? To this I must say no, for these philosophers have put paid to much more naive views still. In so doing they have done useful work in removing defective views, uncovering their mistakes and thus spreading a transition to clearer views.

Similar events often occur in the field of other sciences, for which I cite the example of Wilhelm Weber, who had put forward a theory of electricity and magnetism that is now recognized as incorrect; and yet he is one of those who have most advanced these two sciences. He gave the incentive for many experiments by which the ground was cleared for the newer theories. Although Weber's is today untenable, he is one of the greatest theorists of electricity ever.

From this point of view I must express my sincere thanks to those who recommended my appointment to the lecturership in philosophy, since this has given me the opportunity to delve more deeply into its literature. I cannot judge how many so far have really profited from my lectures but I have this consolation that one at least has learnt a great deal, namely myself.

It is another question whether those who so recommended are satisfied with me. If they expected that I was going to step on the old rails and trundle along them they have indeed deceived themselves. Perhaps this would not even be desirable. Might it not be more useful to have a pike in the carp pond than yet another carp?

In my view all salvation for philosophy may be expected to come from Darwin's theory. As long as people believe in a special spirit that can cognize objects without mechanical means, or in a special will that likewise is apt to will that which is beneficial to us, the simplest psychological phenomena defy explanation.

Only when one admits that spirit and will are not something over and above the body but rather the complicated actions of material parts whose ability so to act becomes increasingly perfected by development, only when one admits that intuition, will and self-consciousness are merely the highest stages of development of those physico-chemical forces of matter by which primeval protoplasmic bubbles were enabled to seek regions that were more and avoid those that were less favourable for them, only then does everything become clear in psychology.

We then understand that with every perception and decision of the will purely mechanical processes are connected, that sensation and will at once act quite perversely and incorrectly if these mechanical processes are disturbed, and that they stop if the disturbance becomes greater still. We further understand that as soon as different ideas enter into mutual interaction, fibres form between the corresponding neurones, that when a child begins to combine visual and auditory sensations fibres form between the brain centres for sight and hearing and similarly between the centres for sight and touch and the motor nerves when it begins to grasp at objects.

We then understand how self-interest is predominant in mankind yet the instinct of self-sacrifice is not entirely lacking. We understand why self-interest must be limited and opposed by laws whereas the instinct of self-sacrifice for the community is promoted in every way by praise and reward. We understand that the innate striving after independence grows to an otherwise quite incomprehensible pertinacity, because weaklings in whom the drive is deficient will go down in the struggle for existence.

Consider another example that is quite simple and banal. Of our original ancestors countless numbers must have died of drinking bad water. Those who preferred the juice of fruit had an advantage. But unfermented fruit juice too could easily contain bacteria, so that those who preferred fermented juices had an advantage in the struggle for existence, and by hereditary development of this predilection it has become a habit which of course often overshoots the mark. I must confess that if I were an antialcoholic I might not have come back alive from America, so severe was the dysentery that I caught there as a result of bad water; even bottles carrying labels of mineral water probably contain mostly river water, and it was only through alcoholic beverages that I was saved.

What then will be the position of the so-called laws of thought in logic? Well, in the light of Darwin's theory they will be nothing else but inherited habits of thought. Men have gradually become accustomed to fix and combine the words through which they communicate and which they rehearse in silence when they think, as well as the memory pictures of those words and everything in the way of internal ideas used for denoting of things, in such a manner as to enable them always to intervene in the world of phenomena in the way intended, and inducing others to do likewise, that is to communicate with them. These interventions are greatly promoted by storing and suitable ordering of memory pictures and by learning and practising speech, and this promotion is the criterion of truth.

This method for putting together and silently rehearsing mental images as well as spoken words became increasingly perfect and has passed into heredity in such a way that fixed laws of thought have developed. It is quite correct that we should not bring these laws with us if all cognition were to cease and perceptions were totally disconnected.

Since therefore the will, that is, the inherited striving towards intervening in the world of phenomena in a way that is helpful to us, has resulted in our ideas becoming gradually perfected, we have indeed the world as will and idea, even Schopenhauer could not wish better.

One can call these laws of thought a priori because through many thousands of years of our species' experience they have become innate to the individual, but it seems to be no more than a logical howler of Kant's to infer their infallibility in all cases.

According to Darwin's theory this howler is perfectly explicable. Only what is certain has become hereditary; what was incorrect has been dropped. In this way these laws of thought acquired such a semblance of infallibility that even experience was believed to be answerable to their judgment. Since they were called a priori, it was concluded that everything a priori was infallible. Just so it was at one time assumed that the ear and the eye also were absolutely perfect because they have indeed developed to an amazing degree of perfection. Today we know that this is an error and that they are not perfect.

Similarly I would deny that our laws of thought are absolutely perfect. On the contrary, they have become such firmly established habits that they overshoot the mark and will not let go even when they are out of place. In this they behave no differently from all inherited habits.

For example, babies have a sucking instinct, otherwise they could not stay alive, and this instinct became so habitual that later the child continues to suck empty rubber. Likewise the laws of thought often overshoot the mark and the philosopher seeks to suck a whole theory of the world out of the concept of nothingness. Likewise the old established and hereditary custom of asking for the cause (the child's eternal question 'why?' already shows it to be hereditary) overshoots the mark if we ask for the cause why the law of cause and effect itself holds; likewise if we ask why the world exists at all, why it is as it is, why we exist at all and why precisely now and so on.

What is particularly striking in all this is the need to ask the question and that the tormenting feeling of not finding an answer does not cease once we have recognized that the framing of the question is in itself misguided. Yet precisely this phenomenon is perfectly explicable on Darwin's theory; habit is simply stronger than recognition that the question is useless. Deceptions of the senses likewise do not cease even when they have been completely explained in terms of physics and physiology. In this way a deception of the intellect supervenes with philosophical problems.

The same holds for the instinct of classification. This is of course something very useful, and one must try to classify as logically as possible. This produces a drive to classify everything, to force everything into a schema like a bed of Procrustes and arbitrarily make things shorter or longer just to fit them into the preconceived idea of the schema.

In this way we take a lot of concepts to be clear or even a priori when they are really mere empty words. We imagine ourselves to be heaven knows how learned if without linking the words in question with clear concepts we ask whether something is synthetic or analytic, transcendental or empirical, real or ideal or material, quantitative or qualitative. About such questions philosophers are apt to write whole treatises; only, whether they are completely clear as to the meaning of their questions, about that they do not ask.

Another example: we are in the habit of assessing everything as to its value; according to whether it helps or hinders the conditions of life, it is valuable or valueless. This becomes so habitual that we imagine we must ask ourselves whether life itself has a value. That is one of those questions utterly devoid of sense. Life itself we must accept as that which has value, and whether something else does can only be judged relatively to life, namely whether it is apt to promote life or not. In this we try of course to talk the individual into believing that what has value for him is not what promotes his own individual life but that of his family, tribe or even mankind as a whole. Since those who believe this (the noble-minded) are in every way furthered and rewarded by the community, they have more opportunities in the struggle for existence and noble-mindedness becomes hereditary, as also unfortunately does self-interest, which for its part offers opportunities in the opposite sense.

However, if we ask whether life itself has value, this means whether life is apt to promote life, a question that has no sense. According to the definition we can ask only how life can be promoted. The valuable is simply what promotes life, the question as to the value of life itself is senseless, although according to Darwin's theory it is readily explicable why it obtrudes itself. It is another mental habit that overshoots the mark.

In a correspondence with Professor Brentano about related questions I once used a comparison which may be trivial but is nevertheless apposite. The wish to produce something when nothing further can be produced I compared first with children eagerly sucking empty rubber, then with the nausea under migraine where one also has the urge to throw up something when nothing is left inside. With this we may compare the attempt to determine whether life has a value, why things are precisely as they are and so on. Grillparzer puts it in a similar way:

> Unto a mill the mind compare,
> That grinds whatever grain is there;
> But if no more of it is brought,
> The stones on one another caught
> Mere splinters, dust and sand prepare.

The task of philosophy for the future is, in my view, to formulate the fundamental concepts in such a way that in all cases we obtain as precise instructions as possible for appropriate interventions in the world of phenomena. This requires first that if we follow different paths we never reach different rules for further thought and action, that is we never meet internal inconsistencies, such as if one path led us to the conclusion that matter was not infinitely divisible and another that it must be. That sort of event is always a sign that the laws of thought still lack the last finish, that we have placed our words badly. In that case we must alter the laws of thought that lead to such absurd consequences.

A similar procedure is followed in algebra. Operations with negative and fractional numbers are defined in such a way that in applying the rules that hold for positive integers we nowhere encounter inconsistencies.

Secondly, the laws of thought must everywhere lead us to the interventions in actual events in the way that experience shows to be desirable.

Thirdly, we must, as best we can, work against the irresistible urge to apply the laws of thought even where they overshoot the mark, so that in the end this urge gradually disappears altogether.

That this is not absolutely impossible history seems to indicate. After all, there was a time when it was believed as a result of logical thought that there could be no antipodes. People always saw that the vertical direction is parallel for everybody and if somebody stands in the opposite sense, his head is on the ground and his feet in the air. By constant experience this becomes so much a habit of thought that the antipodes become inconceivable. People even believed it to be impossible that the Earth was turning, because all other rotations make us dizzy, but not that of the Earth. In the time of Columbus and Copernicus it was believed that this is a necessity of thought, and even they had this foisted on them. However, today these mental habits have vanished and any educated man can now barely comprehend how people could be so narrow-minded then.

The prejudice against non-Euclidean geometry and four-dimensional space is likewise in the course of disappearing. Most people still believe that Euclid's geometry alone is possible, that the sum of the angles in a triangle must be 180°, but some have already come to admit that these are mental ideas become habitual, from which we can and must free ourselves.

Thus we must change all laws of thought in such a way that they lead everywhere to the same goal, that they correspond to experience and that overshooting the mark is kept within proper bounds. Even if this ideal will presumably never be completely realized, we can nevertheless come nearer to it, and this would ensure cessation of the disquiet and the embarrassing feeling that it is a riddle that we are here, that the world is at all and is as it is, that it is incomprehensible what is the cause of this regular connection between cause and effect, and so on. Men would be freed from the spiritual migraine that is called metaphysics.

NOTE

[1] Schopenhauer, *Werke* (Wiesbaden edition) I. 40.

PART II

From *Nature* **51** (1895)

ON CERTAIN QUESTIONS OF THE
THEORY OF GASES*

§1. I propose to answer two questions: (1) Is the Theory of Gases a true physical theory as valuable as any other physical theory? (2) What can we demand from any physical theory?

The first question I answer in the affirmative, but the second belongs not so much to ordinary physics (let us call it orthophysics) as to what we call in Germany metaphysics. For a long time the celebrated theory of Boscovich was the ideal of physicists. According to his theory, bodies as well as the ether are aggregates of material points, acting together with forces, which are simple functions of their distances. If this theory were to hold good for all phenomena, we should be still a long way off what Faust's famulus hoped to attain, viz. to know everything. But the difficulty of enumerating all the material points of the universe, and of determining the law of mutual force for each pair, would be only a quantitative one; nature would be a difficult problem, but not a mystery for the human mind.

When Lord Salisbury says that nature is a mystery[1], he means, it seems to me, that this simple conception of Boscovich is refuted almost in every branch of science, the Theory of Gases not excepted. The assumption that the gas-molecules are aggregates of material points, in the sense of Boscovich, does not agree with the facts. But what else are they? And what is the ether through which they move? Let us again hear Lord Salisbury. He says:

What the atom of each element is, whether it is a movement, or a thing, or a vortex, or a point having inertia, all these questions are surrounded by profound darkness. I dare not use any less pedantic word than entity to designate the ether, for it would be a great exaggeration of our knowledge if I were to speak of it as a body, or even as a substance.

If this be so – and hardly any physicist will contradict this – then neither the Theory of Gases nor any other physical theory can be quite a con-

* *Nature* **51** (1895) 413–415. Reprinted in *WA* III, 535–544.

gruent account of facts, and I cannot hope with Mr Burbury, that Mr Bryan will be able to deduce all the phenomena of spectroscopy from the electromagnetic theory of light. Certainly, therefore, Hertz is right when he says:[2] "The rigour of science requires that we distinguish well the undraped figure of nature itself from the gay-coloured vesture with which we clothe it at our pleasure." But I think the predilection for nudity would be carried too far if we were to forgo every hypothesis. Only we must not demand too much from hypotheses.

It is curious to see that in Germany, where till lately the theory of action at a distance was much more cultivated than in Newton's native land itself, where Maxwell's theory of electricity was not accepted, because it does not start from quite a precise hypothesis, at present every special theory is old-fashioned, while in England interest in the Theory of Gases is still active; vide, among others, the excellent papers of Mr Tait, of whose ingenious results I cannot speak too highly, though I have been forced to oppose them in certain points.

Every hypothesis must derive indubitable results from mechanically well-defined assumptions by mathematically correct methods. If the results agree with a large series of facts, we must be content, even if the true nature of facts is not revealed in every respect. No one hypothesis has hitherto attained this last end, the Theory of Gases not excepted. But this theory agrees in so many respects with the facts, that we can hardly doubt that in gases certain entities, the number and size of which can roughly be determined, fly about pell-mell. Can it be seriously expected that they will behave exactly as aggregates of Newtonian centres of force, or as the rigid bodies of our Mechanics? And how awkward is the human mind in divining the nature of things, when forsaken by the analogy of what we see and touch directly?

The following assumptions, while not professing to explain the mysteries to which Lord Salisbury alluded, nevertheless show that it is possible to explain the spectra of gases while ascribing 5 degrees of freedom to the molecules, and without departing from Boscovich's standpoint.

Let the molecules of certain gases behave as rigid bodies. The molecules of the gas and of the enclosing vessel move through the ether without loss of energy as rigid bodies, or as Lord Kelvin's vortex rings move through a frictionless liquid in ordinary hydrodynamics. If we were to take a vessel filled with one gram of gas kept during an infinitely long time always at 0°C. and containing always the same portion of ether, every atom of our gas molecules would reach the same average *vis viva*. If then we were to raise the temperature to 1°C. and to wait till every ponderable and every ether atom was in thermal equilibrium, the total energy would be augmented by what we may call the ideal specific heat. But in actually heating one gram of gas, the ether always flows freely through the walls of the vessel. It comes from the universe, and is not at all in thermal equilibrium with the molecules of the gas. It is true that it always carries off energy, if the outside space is colder than the gas; but this energy may be so small as to be quite negligible in comparison with the energy which the gas loses by heat-conduction, and which must be experimentally determined and subtracted in measuring the specific heat. Only certain transverse vibrations of the ether can transfer sensible energy from one ponderable body to another, and therefore a correction for radiant heat must be applied to observations of specific heats. These transverse vibrations are not produced (as in the older theories of light) by simple atomic vibrations, but their pitch depends on the shape of the hollow space which the molecule forms in the ether, just as Hertzian waves are not caused by vibrations of the ponderable matter of the brass balls, the form of which only determines the pitch. The unknown electric action accompanying a chemical process augments these transverse vibrations enormously. The generalised co-ordinates of the ether, on which these vibrations depend, have not the same *vis viva* as the coordinates which determine the position of a molecule, because the entire ether has not had time to come into thermal equilibrium with the gas molecules, and has in no respect attained the state which it would have if it were enclosed for an infinitely long time in the same vessel with the molecules of the gas.

But how can the molecules of a gas behave as rigid bodies? Are they not composed of smaller atoms? Probably they are; but the *vis viva* of their internal vibrations is transformed into progressive and rotatory motion so slowly, that when a gas is brought to a lower temperature the molecules may retain for days, or even for years, the higher *vis viva* of their internal vibrations corresponding to the original temperature. This transference of energy, in fact, takes place so slowly that it cannot be perceived amid the fluctuations of temperature of the surrounding bodies.

The possibility of the transference of energy being so gradual cannot be denied, if we also attribute to the ether so little friction that the Earth is not sensibly retarded by moving through it for many hundreds of years.

If the ether be an external medium which flows freely through the gas, we might find a difficulty in explaining how it is that the source of radiant heat seems to be in the energy of the gas itself. But I still think it possible that the source of energy of the electric vibrations caused by the impact of two gas molecules in the surrounding ether, may be in the progressive and rotatory energy of the molecule. If the electric states of two molecules differ in their motions of approach and separation, the energy of progressive motion may be transformed into electric energy.

Moreover, it is doubtful whether emission of rays of visible light takes place in simple gases without chemical action. Certainly the light of sodium and that of Gassiot's tubes do not come from gases whose molecules are in thermal equilibrium.

It may be objected that the above is nothing more than a series of imperfectly proved hypotheses. But granting its improbability, it suffices that this explanation is not impossible. For then I have shown that the problem is not insoluble, and nature will have found a better solution than mine.

§2. Mr Culverwell's objections[3] against my Minimum Theorem bear the closest connections to what I pointed out in the second part of my paper 'Bemerkungen über einige Probleme der mechanischen Wärmetheorie'[4]. There I pointed out that my Minimum Theorem, as well as the so-called Second Law of Thermodynamics, are only theorems of probability. The Second Law can never be proved mathematically by means of the equations of dynamics alone.

Let us compare two motions of a dynamical system. At the beginning of the second motion, let the coordinates specifying the position of every part of the moving system, and the magnitudes of all the corresponding velocities, be the same as they were at the end of the first motion, but let the direction of every velocity be exactly reversed. Then in the second motion the system moves exactly in the opposite way to what it does in the first; hence, if for the first motion we have

$$\int \frac{dQ}{T} < 0,$$

then for the second we must have

$$\int \frac{dQ}{T} > 0.$$

That is, if under certain conditions

$$\int \frac{dQ}{T} < 0,$$

we can always find other initial conditions which give to the same system with the same equations of motion,

$$\int \frac{dQ}{T} > 0.$$

In the same manner, Mr Culverwell wishes to refute my Minimum Theorem. Mr. Culverwell's reasoning is suspicious, because by the same reasoning we could prove that oxygen and nitrogen do not diffuse. Suppose that initially one half of a closed vessel contains pure oxygen, and in the other half pure nitrogen; when the diffusion has advanced for a certain time, reverse the directions of all velocities, then the gases separate again, and, according to Mr Culverwell's argument, we could believe that the probability that oxygen and nitrogen separate is as great as the probability that they mix.

Though interesting and striking at the first moment, Mr Culverwell's arguments rest, as I think, only upon a mistake of my assumptions. It can never be proved from the equations of motion alone, that the minimum function H must always decrease. It can only be deduced from the laws of probability that if the initial state is not specially arranged for a certain purpose, but haphazard governs freely, the probability that H decreases is always greater than that it increases. It is well known that the theory of probability is as exact as any other mathematical theory, if properly understood. If we make 6000 throws with dice, we cannot prove that we shall throw any particular number exactly 1000 times; but we can prove that the ratio of the number of throws, approaches the more to $\frac{1}{6}$ the oftener we throw.

Let us now take a given rigid vessel with perfectly smooth and perfectly elastic walls containing a given number of gas-molecules moving for an indefinitely long time. All *regular* motions (e.g. one where all the molecules move in one plane) shall be excluded. During the greater part of this time H will be very nearly equal to its minimum value H (min.). Let us construct the H-curve, i.e. let us take the time as axis of abscissae and draw the curve, whose ordinates are the corresponding values of H. The greater majority of the ordinates of this curve are very nearly equal to H (min.). But because greater values of H are not mathematically impossible, but only very improbable, the curve has certain, though very few, summits or maximum ordinates which rise to a greater height than H (min.).

We will now consider a certain ordinate $H_1 > H$ (min.). Two cases are possible. H_1 may be very near the top of a summit, so that H decreases if we go either in the positive or negative direction along the axis representing time. The second case is that H_1 lies in a part of the curve ascending to or descending from a higher summit. Then the ordinates on the one side of H_1 will be greater, and on the other less than H_1. But because higher summits are so extremely improbable, the first case will be the most probable, and if we choose an ordinate of given magnitude H_1 guided by haphazard in the curve, it will be not certain, but very probable, that the ordinate decreases if we go in either direction.

We will now assume, with Mr Culverwell, a gas in a given state. If in this state H is greater than H (min.) it will be not certain, but very probable, that H decreases and finally reaches not exactly but very nearly the value H (min.), and the same is true at all subsequent instants of time. If in an intermediate state we reverse all velocities, we get an exceptional case, where H increases for a certain time and then decreases again. But the existence of such cases does not disprove our theorem. On the contrary, the theory of probability itself shows that the probability of such cases is not mathematically zero, only extremely small.

Hence Mr Burbury is wrong, if he concedes that H increases in as many cases as it decreases, and Mr Culverwell is also wrong, if he says that all that any proof can show is that taking all values of dH/dt got from taking all the configurations which approach towards a permanent state, and all the configurations which recede from it, and then striking some average, dH/dt would be negative. On the contrary, we have shown the possibility that H may have a tendency to decrease, whether we pass to the former or to the latter configurations. What I proved in my papers is as follows: It is extremely probable that H is very near to its minimum value; if it is greater, it may increase or decrease, but the probability that it decreases is always greater. Thus, if I obtain a certain value for dH/dt, this result does not hold for every time-element dt, but is only an average value. But the greater the number of molecules, the smaller is the time-interval dt for which the results holds good.

I will not here repeat the proofs given in my papers; I will only show that just the same takes place in the much simpler case of dice. We will make an indefinitely long series of throws with a die. Let A_1 be the number of times of throwing the number 1, among the first $6n$ throws, A_2 the number of times of throwing 1, among all the throws between the second and the $(6n+1)$th inclusive, and so on. Let us construct a series of points in a plane, the successive abscissae of which are

$$0, \ \frac{1}{n}, \ \frac{2}{n}, \ \frac{3}{n}, \dots,$$

the ordinates of which are

$$y_1 = \left(\frac{A_1}{n} - 1\right)^2, \quad y_2 = \left(\frac{A_2}{n} - 1\right)^2 \dots;$$

let us call this series of points the 'P-curve'. If n is a large number, the greater proportion of the ordinates of this new curve will be very small. But the P-curve (like the afore-mentioned H-curve) has summits which are higher than the ordinary course of the curve. Let us now consider all the points of the P-curve, whose ordinates are exactly $= 1$. We will call these points 'the points B'. Since for each point $y = (A/n - 1)^2$, therefore for the point B we have $A = 2n$; these points mark, therefore, the case where, by chance, we have thrown the number 1 in $2n$ out of $6n$ throws. If n is at all large, that is extremely improbable, but never absolutely impossible. Let v be a number much smaller than n, and let us go forward from the abscissa of each point B through a distance $= 6v/n$ in the direction of x positive. We shall probably meet a point, the ordinate of which < 1. The probability that we meet an ordinate > 1 is extremely small, but not zero. By reasoning in the same manner as Mr Culverwell, we might believe that if we go backward (i.e. in the direction of x negative) from the abscissa of each point B through a distance $= 6v/n$, it would be probable that we should meet ordinates > 1. But this inference is not correct. Whether we go in the positive or in the negative direction the ordinates will probably decrease.

We can even calculate the probable diminution of y. We have seen that for every point B we have $A = 2n$ (i.e. $2n$ throws out of $6n$ turning up 1). If we move in the positive or negative direction along the axis of x through the distance $1/n$, we exclude one of the $6n$ throws, and we include a new one. When we move forward through the distance $6v/n$, we have excluded $6v$ of the original throws, and included $6v$ others. Among the excluded throws we have probably $2v$, among the included ones v throws of the number 1. Therefore the probable diminution of A is v, the probable diminution of y is $2v/n$ approximately. Because the variation of x was $6v/n$, we may write

$$\frac{dy}{dx} = -\tfrac{1}{3}.$$

But this is not an ordinary differential coefficient. It is only the average ratio of the increase of y to the corresponding increase of x for all points, whose ordinates are $= 1$. The P-curve belongs to the large class of curves which have nowhere a uniquely defined tangent. Even at the top of each summit the tangent is not parallel to the x-axis, but is undefined. In other words, the chord joining two points of the curve does not tend toward a definite limiting position when one of the two points approaches and ultimately coincides with the other.[5] The same applies to the H-curve in the Theory of Gases. If I find a certain negative value for dH/dt that does not define the tangent of the curve in the ordinary sense, but it is only an average value.

§3. Mr Culverwell says that my theorem cannot be true because if it were true every atom of the universe would have the same average *vis viva*, and all energy would be dissipated. I find, on the contrary, that this argument only tends to confirm my theorem, which requires only that in the course of time the universe must tend to a state where the average *vis viva* of every atom is the same and all energy is dissipated, and that is indeed the case. But if we ask why this state is not yet reached, we again come to a 'Salisburian mystery'.

I will conclude this paper with an idea of my old assistant, Dr Schuetz. We assume that the whole universe is, and rests for ever, in thermal equilibrium. The probability that one (only one) part of the universe is in a certain state, is the smaller the further this state is from thermal equilibrium; but this probability is greater, the greater the universe itself is. If we assume the universe great enough we can make the probability of one relatively small part being in any given state (however far from the state of thermal equilibrium), as great as we please. We can also make the probability great that, though the whole universe is in thermal equilibrium, our world is in its present state. It may be said that the world is so far from thermal equilibrium that we cannot imagine the improbability of such a state. But can we imagine, on the other side, how small a part of the whole universe this world is? Assuming the universe great enough, the probability that such a small part of it as our world should be in its present state, is no longer small.

If this assumption were correct, our world would return more and more to thermal equilibrium; but because the whole universe is so great, it might be probable that at some future time some other world might deviate as far from thermal equilibrium as our world does at present. Then the aforementioned H-curve would form a representation of what takes place in the universe. The summits of the curve would represent the worlds where visible motion and life exist.

NOTES

1 Presidential Address to the British Association at Oxford, 1894.
2 Hertz, 'Untersuchungen über die Ausbreitung der elektrischen Kraft' (Barth, Leipzig 1892, p.31).
3 *Editor's note*: In *Nature* 50 (1894) 617, E. P. Culverwell initiated a correspondence about Boltzmann's kinetic theory of gases, in which S. H. Burbury, G. H. Bryan, H. W. Watson, and Boltzmann himself (by the present paper and by one other contribution) took part. See *Nature* 50–51.
4 *Wien. Ber.* 75, 1877 = *WA* II, 112–148.
5 See Ulisse Dini, *Grundlagen für eine Theorie der Funktionen einer reellen Veränderlichen* (Teubner 1892, §126), or Weierstrass, *Journal für die Mathematik* 79, p. 29.

PART III

From *Encyclopaedia Britannica*[10,11]

MODEL*

Model (O. Fr. *modelle*, mod. *modèle*; It. *modello*, pattern, mould; from Lat. *modus*, measure, standard), a tangible representation, whether the size be equal, or greater, or smaller, of an object which is either in actual existence, or has to be constructed in fact or in thought. More generally it denotes a thing, whether actually existing or only mentally conceived of, whose properties are to be copied. In foundries, the object of which a cast is to be taken, whether it be for engineering or artistic purposes, is usually first formed of some easily workable material, generally wood. The form of this model is then reproduced in clay or plaster, and into the mould thus obtained the molten metal is poured. The sculptor first makes a model of the object he wishes to chisel in some plastic material such as wax, ingenious and complicated contrivances being employed to transfer this wax model, true to nature, to the stone in which the final work is to be executed. In anatomy and physiology, models are specially employed as aids in teaching and study, and the method of *moulage* or chromoplastic yields excellent impressions of living organisms, and enables anatomical and medical preparations to be copied both in form and colour. A special method is also in use for making plastic models of microscopic and minute microscopic objects. That their internal nature and structure may be more readily studied, these are divided by numerous parallel transverse cuts, by means of a microtome, into exceedingly thin sections. Each of these shavings is then modelled on an enlarged scale in wax or pulp plates, which are fixed together to form a reproduction of the object.

Models in the mathematical, physical and mechanical sciences are of the greatest importance. Long ago philosophy perceived the essence of our process of thought to lie in the fact that we attach to the various real objects around us particular physical attributes – our concepts – and by means of these try to represent the objects to our minds. Such views were formerly regarded by mathematicians and physicists as nothing more than unfertile speculations, but in more recent times they have been

* First published 1902, here reprinted from the edition of 1910–11.

214 FROM 'ENCYCLOPAEDIA BRITANNICA'

brought by J. C. Maxwell, H. v. Helmholtz, E. Mach, H. Hertz and many others into intimate relation with the whole body of mathematical and physical theory. On this view our thoughts stand to things in the same relation as models to the objects they represent. The essence of the process is the attachment of one concept having a definite content to each thing, but without implying complete similarity between thing and thought; for naturally we can know but little of the resemblance of our thoughts to the things to which we attach them. What resemblance there is lies principally in the nature of the connexion, the correlation being analogous to that which obtains between thought and language, language and writing. the notes on the stave and musical sounds, &c. Here, of course, the symbolization of the thing is the important point, though, where feasible, the utmost possible correspondence is sought between the two – the musical scale, for example, being imitated by placing the notes higher or lower. When, therefore, we endeavour to assist our conceptions of space by figures, by the methods of descriptive geometry, and by various thread and object models; our topography by plans, charts and globes; and our mechanical and physical ideas by kinematic models – we are simply extending and continuing the principle by means of which we comprehend objects in thought and represent them in language or writing. In precisely the same way the microscope or telescope forms a continuation and multiplication of the lenses of the eye; and the notebook represents an external expansion of the same process which the memory brings about by purely internal means. There is also an obvious parallelism with representation by means of models when we express longitude, mileage, temperature, &c., by numbers, which should be looked upon as arithmetical analogies. Of a kindred character is the representation of distances by straight lines, of the course of events in time by curves, &c. Still, neither in this case nor in that of maps, charts, musical notes, figures, &c., can we legitimately speak of models, for these always involve a concrete spatial analogy in three dimensions.

So long as the volume of matter to be dealt with in science was insignificant, the need for the employment of models was naturally less imperative; indeed, there are self-evident advantages in comprehending things without resort to complicated models, which are difficult to make, and cannot be altered and adapted to extremely varied conditions so readily as can the easily adjusted symbols of thought, conception and calculation. Yet as the facts of science increased in number, the greatest economy of effort had to be observed in comprehending them and in conveying them to others; and the firm establishment of ocular demonstration was inevitable in view of its enormous superiority over purely abstract symbolism for the rapid and complete exhibition of complicated relations. At the present time it is desirable, on the one hand, that the power of deducing results from purely abstract premisses, without recourse to the aid of tangible models, should be more and more perfected, and on the other that purely abstract conceptions should be helped by objective and comprehensive models in cases where the mass of matter cannot be adequately dealt with directly.

In pure mathematics, especially geometry, models constructed of papier mâché and plaster are chiefly employed to present to the senses the precise form of geometrical figures, surfaces and curves. Surfaces of the second order, represented by equations of the second degree between the rectangular co-ordinates of a point, are very simple to classify, and accordingly all their possible forms can easily be shown by a few models, which, however, became somewhat more intricate when lines of curvature, loxodromics and geodesic lines have to appear on their surfaces. On the other hand, the multiplicity of surfaces of the third order is enormous, and to convey their fundamental types it is necessary to employ numerous models of complicated, not to say hazardous, construction. In the case of more intricate surfaces it is sufficient to present those singularities which exhibit variation from the usual type of surface with synclastic or anticlastic curvatures, such as, for example, a sharp edge or point, or an intersection of the surface with itself; the elucidation of such singularities is of fundamental importance in modern mathematics.

In physical science, again, models that are of unchangeable form are largely employed. For example, the operation of the refraction of light in crystals can be pictured if we imagine a point in the centre of the crystal whence light is dispersed in all directions. The aggregate of the places at which the light arrives at any instant after it has started is called the wavefront. This surface consists of two cups or sheets fitting closely and exactly one inside the other. The two rays into which a single ray is broken are always determined by the points of contact of certain tangent-planes drawn to those sheets. With crystals possessing two axes these wave-surfaces display peculiar singularities in the above sense of the term, in that the inner sheet has four protuberances, while the outer has four funnel-like depressions, the lowest point of each depression meeting the highest point of each protuberance. At each of these funnels there is a tangent-plane that touches not in a single point, but in a circle bounding the depression, so that the corresponding ray of light is refracted, not into two rays, but into a whole cone of light – the so-called conical refraction theoretically predicted by Sir W. R. Hamilton and experimentally detected by Humphrey Lloyd. These conditions, which it is difficult to adequately express in language, are self-evident so soon as the wave-surface formed in plaster lies before our eyes. In thermodynamics, again, similar models serve, among other purposes, for the representation of the surfaces which exhibits the relation between the three thermodynamic variables of a body, e.g. between its temperature, pressure and volume. A glance at the model of such a thermodynamic surface enables the behaviour of a particular substance under the most varied conditions to be immediately realized. When the ordinate intersects the surface but once a single phase only of the body is conceivable, but where there is a multiple intersection various phases are possible, which may be liquid or gaseous. On the boundaries between these regions lie the critical phases, where transition occurs from one type of phase into the other. If for one of the elements a quantity which occurs in calorimetry be chosen – for example, entropy – information is also gained about the behaviour of the body when heat is taken in or abstracted.

After the stationary models hitherto considered, come the manifold forms of moving models, such as are used in geometry, to show the origin of geometrical figures from the motion of others – e.g. the origin of surfaces from the motion of lines. These include the thread models, in which threads are drawn tightly between movable bars, cords, wheels, rollers, &c. In mechanics and engineering an endless variety of working models are employed to convey to the eye the working either of machines as a whole, or of their component and subordinate parts. In theoretical mechanics models are often used to exhibit the physical laws of motion in interesting or special cases e.g. the motion of a falling body or of a spinning-top, the movement of a pendulum on the rotating earth, the vortical motions of fluids, &c. Akin to these are the models which execute more or less exactly the hypothetical motions by which it is sought to explain various physical phenomena – as, for instance, the

complicated wave-machines which present the motion of the particles
in waves of sound (now ascertained with fair accuracy), or the more
hypothetical motion of the atoms of the aether in waves of light.

The varying importance which in recent times has been attached to
models of this kind is intimately connected with the changes which have
taken place in our conceptions of nature. The first method by which an
attempt was made to solve the problem of the universe was entirely under
the influence of Newton's laws. In analogy to his laws of universal
gravitation, all bodies were conceived of as consisting of points of matter –
atoms or molecules – to which was attributed a direct action at a distance.
The circumstances of this action at a distance, however, were conceived
as differing from those of the Newtonian law of attraction, in that they
could explain the properties not only of solid elastic bodies, but also
those of fluids, both liquids and gases. The phenomena of heat were
explained by the motion of minute particles absolutely invisible to the
eye, while to explain those of light it was assumed that an impalpable
medium, called luminiferous aether, permeated the whole universe;
to this were attributed the same properties as were possessed by solid
bodies, and it was also supposed to consist of atoms, although of a
much finer composition. To explain electric and magnetic phenomena
the assumption was made of a third species of matter – electric fluids
which were conceived of as being more of the nature of fluids, but still
consisting of infinitesimal particles, also acting directly upon one another
at a distance. This first phase of theoretical physics may be called the
direct one, in that it took as its principal object the investigation of the
internal structure of matter as it actually exists. It is also known as the
mechanical theory of nature, in that it seeks to trace back all natural
phenomena to motions of infinitesimal particles, i.e. to purely mechanical
phenomena. In explaining magnetic and electrical phenomena it inevitably
fell into somewhat artificial and improbable hypotheses, and this induced
J. Clerk Maxwell, adopting the ideas of Michael Faraday, to propound
a theory of electric and magnetic phenomena which was not only new in
substance, but also essentially different in form. If the molecules and
atoms of the old theory were not to be conceived of as exact mathematical
points in the abstract sense, then their true nature and form must be
regarded as absolutely unknown, and their groupings and motions,
required by theory, looked upon as simply a process having more or less

resemblance to the workings of nature, and representing more or less exactly certain aspects incidental to them. With this in mind, Maxwell propounded certain physical theories which were purely mechanical so far as they proceeded from a conception of purely mechanical processes. But he explicitly stated that he did not believe in the existence in nature of mechanical agents so constituted, and that he regarded them merely as means by which phenomena could be reproduced, bearing a certain similarity to those actually existing, and which also served to include larger groups of phenomena in a uniform manner and to determine the relations that held in their case. The question no longer being one of ascertaining the actual internal structure of matter, many mechanical analogies or dynamical illustrations became available, possessing different advantages; and as a matter of fact Maxwell at first employed special and intricate mechanical arrangements, though later these became more general and indefinite. This theory, which is called that of mechanical analogies, leads to the construction of numerous mechanical models. Maxwell himself and his followers devised many kinematic models, designed to afford a representation of the mechanical construction of the ether as a whole as well as of the separate mechanisms at work in it: these resemble the old wave-machines, so far as they represent the movements of a purely hypothetical mechanism. But while it was formerly believed that it was allowable to assume with a great show of probability the actual existence of such mechanisms in nature, yet nowadays philosophers postulate no more than a partial resemblance between the phenomena visible in such mechanisms and those which appear in nature. Here again it is perfectly clear that these models of wood, metal and cardboard are really a continuation and integration of our process of thought; for, according to the view in question, physical theory is merely a mental construction of mechanical models, the working of which we make plain to ourselves by the analogy of mechanisms we hold in our hands, and which have so much in common with natural phenomena as to help our comprehension of the latter.

Although Maxwell gave up the idea of making a precise investigation into the final structure of matter as it actually is, yet in Germany his work, under G. R. Kirchhoff's lead, was carried still further. Kirchhoff defined his own aim as being to describe, not to explain, the world of phenomena; but as he leaves the means of description open his theory differs little from Maxwell's, so soon as recourse is had to description by means of mechanical models and analogies. Now the resources of pure mathematics being particularly suited for the exact description of relations of quantity, Kirchhoff's school laid great stress on description by mathematical expressions and formulae, and the aim of physical theory came to be regarded as mainly the construction of formulae by which phenomena in the various branches of physics should be determined with the greatest approximation to the reality. This view of the nature of physical theory is known as mathematical phenomenology; it is a presentation of phenomena by analogies, though only by such as may be called mathematical.

Another phenomenology in the widest sense of the term, maintained especially by E. Mach, gives less prominence to mathematics, but considers the view that the phenomena of motion are essentially more fundamental than all the others to have been too hastily taken. It rather emphasizes the prime importance of description in the most general terms of the various spheres of phenomena, and holds that in each sphere its own fundamental law and the notions derived from this must be employed. Analogies and elucidations of one sphere by another – e.g. heat, electricity, &c. – by mechanical conceptions, this theory regards as mere ephemeral aids to perception, which are necessitated by historical development, but which in course of time either give place to others or entirely vanish from the domain of science.

All these theories are opposed by one called energetics (in the narrower sense), which looks upon the conception of energy, not that of matter, as the fundamental notion of all scientific investigation. It is in the main based on the similarities energy displays in its various spheres of action, but at the same time it takes its stand upon an interpretation or explanation of natural phenomena by analogies which, however, are not mechanical, but deal with the behaviour of energy in its various modes of manifestation.

A distinction must be observed between the models which have been described and those experimental models which present on a small scale a machine that is subsequently to be completed on a larger, so as to afford a trial of its capabilities. Here it must be noted that a mere alteration in dimensions is often sufficient to cause a material alteration in the action, since the various capabilities depend in various ways on the linear dimensions. Thus the weight varies as the cube of the linear dimensions, the surface of any single part and the phenomena that depend on such surfaces are proportionate to the square, while other effects – such as friction, expansion and condition of heat, &c., vary according to other laws. Hence a flying-machine, which when made on a small scale is able to support its own weight, loses its power when its dimensions are increased. The theory, initiated by Sir Isaac Newton, of the dependence of various effects on the linear dimensions, is treated in the article UNITS, DIMENSIONS OF.* Under simple conditions it may often be affirmed that in comparison with a large machine a small one has the same capacity, with reference to a standard of time which must be diminished in a certain ratio.

Of course experimental models are not only those in which purely mechanical forces are employed, but also include models of thermal, electro-magnetic and other engines – e.g. dynamos and telegraphic machines. The largest collection of such models is to be found in the museum of the Washington Patent Office. Sometimes, again, other than purely mechanical forces are at work in models for purposes of investigation and instruction. It often happens that a series of natural processes – such as motion in liquids, internal friction of gases, and the conduction of heat and electricity in metals – may be expressed by the same differential equations; and it is frequently possible to follow by means of measurements one of the processes in question – e.g. the conduction of electricity just mentioned. If then there be shown in a model a particular case of electrical conduction in which the same conditions at the boundary hold as in a problem of the internal friction of gases, we are able by measuring the electrical conduction in the model to determine at once the numerical data which obtain for the analogous case of internal friction, and which could only be ascertained otherwise by intricate calculations. Intricate calculations, moreover, can very often be dispensed with by the aid of mechanical devices, such as the ingenious calculating machines which perform additions and subtractions and very elaborate multiplications and divisions with surprising speed and accuracy, or apparatus for solving the higher equations, for determining the volume or area of geometrical figures, for carrying out integrations, and for developing a function in a Fourier's series by mechanical means.

(L.Bo.).

* *Editor's note*: EB[11] **27**, 736 ff., an article by Sir J. Larmor.

PART IV

From *Vorlesungen über die Principe der Mechanik*
(*Lectures on the Principles of Mechanics*)

LECTURES ON THE PRINCIPLES OF MECHANICS*

PART I

PREFACE

> Put forward what is true
> So write that it may be clear,
> Fight for it to the end.

IF AT PRESENT I publish not Part Two of my Gas Theory, but Part One of my Mechanics, I would not excuse this by pointing to famous paradigms of the sequence in which volumes appear. The circumstances are rather these: in Part Two of Gas Theory there was such an accumulation of necessary insertions on mechanics that they seemed first to fill a whole paragraph and then a section, so that in the end I decided to make a whole book of it, by adding a further notebook which during the previous holidays I had worked out into a set of lectures on mechanics for the following winter semester. When, however, I considered my audience, it seemed to me that I should exchange the entire method for a simpler one. So that my efforts should not be entirely lost, I absorbed the contents of that notebook into the present book, which I could thus call lectures that I did not give at the University of Vienna.

In recent times there has been much talk about obscurities in the principles of mechanics and attempts have been made to remove them by clothing mechanics in entirely new and alien garb. Here I have chosen the opposite path, trying whether one might not avoid these obscurities while representing mechanics as nearly as possible in its classical form, partly by detailed treatment of certain matters that used to be skipped or merely superficially touched on as being self-evident, partly by taking into account all justified criticisms. Here I warmly agree with Hertz's

* Vorlesungen über die Principe der Mechanik, Part I published 1897, Part II published 1904.

comments on the wealth of ideas in the relevant writings of Mach, even if I am by no means everywhere of the same opinion as Mach.

I should have liked to adopt the method of representation by quaternions, but this would have run counter to my endeavours to exclude everything that was unfamiliar to the German public.

Part Two of the lectures on mechanics which will first mention the principles important for gas theory, is to appear very shortly and next, as soon as I can manage it, Part Two of Gas Theory. A third part of mechanics is to cover elasticity and hydrodynamics, thus leading back to gas theory.

In so extensive and much worked a field as mechanics there can of course be no question of completeness or essential novelty. Nevertheless it became clear from the section on the parallelogram of forces to the definition of the equilibrium of a system, d'Alembert's principle and the general form of the equations of motion, that many special theorems of mechanics still need to be made more strictly precise. Of course none of these questions could be definitely resolved here, that would be possible only in a monograph, but I am content if I have pointed out the gaps and given inspiration for further research.

Abbazia, 3 August, 1897

I. FUNDAMENTAL CONCEPTS

§1. *Characterization of the Method Chosen*

Mechanics is the theory of the motion of natural bodies, that is change of place (relative change of position) that is not connected with any change of their other properties. According to this definition mechanics must also explore under what conditions a body does not change its place; that is, is at rest.

Since changes of place are the simplest phenomena, mechanics is the foundation of all natural science. Small wonder therefore if it early became well developed (by Newton, Lagrange, Euler and so on) and is currently being brought to ever greater degrees of perfection. This perfection however rests more on the certainty with which mechanics treats special problems than on its fundamental principles. The latter have been often attacked especially in quite recent times.

Let me merely cite the famous book by Hertz[1] who does of course admit in the end that the unclarity of the fundamental principles of mechanics is the fault mainly of defective presentation in text books. As to my own view about the cause leading to this defective mode of presentation, I shall shortly come to it.

Nobody surely ever doubted what Hertz emphasizes in his book, namely that our thoughts are mere pictures of objects (or better, signs for them), which at most have some sort of affinity with them but never coincide with them but are related to them as letters to spoken sounds or written notes to musical sounds. Because of our limited intellects these pictures can never reflect more than a small part of objects.

We can now proceed in one of two ways. The first is to leave the pictures more general, so that we run less risk of their later turning out incorrect since they will be more adaptable to new factual findings; however their generality makes these pictures more indefinite and vague and their further development will be connected with some measure of uncertainty and ambiguity. The second is to specialise the pictures and elaborate them to a measure of detail, in which case we shall have to import much more that is arbitrary (hypothetical) and might not fit new experience; but we shall have the advantage that the pictures are as clear and definite as possible so that we can draw from them all consequences fully defined and unambiguous.

It is precisely the unclarities in the principles of mechanics that seem to me to derive from not starting at once with hypothetical mental pictures but trying to link up with experience from the outset. As to the transition to hypotheses, people attempted more or less to conceal it or even to contrive an artificial demonstration that the whole edifice was necessary and free from hypotheses; but just this produced unclarity.

As regards our own times, one often finds the demand that only directly given phenomena should be encompassed and nothing arbitrary added to them. Advisable as it is to separate the factual from the hypothetical and never to multiply the latter beyond need, I believe that without any hypothetical features one could never go beyond an unsimplified memory mark for each separate phenomenon. All simplifications of memory pictures, all laying hold of law-like features, all rules for summarizing complicated phenomena and predetermining them according to simple prescriptions, rest on the use of pictures drawn from other kinds of simple phenomena and acts of the will.

People have put forward as ideal the mere setting up of partial differential equations and prediction of phenomena from them. However, they too are nothing more than rules for constructing alien mental pictures, namely of series of numbers. Partial differential equations require the construction of collections of numbers representing a manifold of several dimensions. If we remember the meaning of their symbolism they are nothing more than the demand to imagine very many points of such manifolds (that is, positions that are characterized by several numbers of the manifold, as spatial points are by their co-ordinates) and, using certain rules, constantly to derive from them new points of the manifold, to imagine, as it were, a progressive movement of the points in the manifold.[2]

Thus if we go to the bottom of it, Maxwell's electromagnetic equations in their Hertzian form likewise contain hypothetical features added to experience, which are fashioned, as always, by transferring the laws we have observed in finite bodies to fictitious elements of our own making. These equations, like all partial differential equations of mathematical physics, which in the case of simultaneous action of several natural forces (electricity, magnetism, elasticity, heat, chemical forces) are almost unimaginably complicated, are likewise only inexact schematic pictures for definite areas of fact, even though the pictures are pieced together from elements that are somewhat different from the atoms to which we are accustomed. The justification of these equations Hertz seeks only after the event in the agreement with experience, just as we should with atomist pictures.

The assertion that atomism does, while partial differential equations do not, introduce material extraneous to the facts seems to me unfounded. Of course we must not infer from the applicability of atomism, so often suggested by the facts, that its pictures must be sufficient everywhere. Where only a strained application of it has been possible, one should adduce other pictures that start from other elements. One is to apply only such atomist pictures as are well-founded in the facts themselves but never do violence to nature by means of arbitrary and fantastic ideas.

As to that, surely nobody will make atomism responsible for all the phantasms perpetrated in its field by the incompetent. Who knows whether energetics would be less afflicted by such excrescences by the time it reached the age of atomism.

Nor must one ever seek metaphysical reasons for the picture nor draw hasty inferences from it, for example that chemical atoms are material points. Nor should we lose from sight the possibility that it might one day be displaced by quite different pictures, let us say, to avoid appearing small-hearted, ones taken from manifolds that lack even the properties of our three-dimensional space, so that for example simple geometrical constructions of atomism would have to be replaced by manipulations with numbers forming a complicated manifold.

I am thus the last to deny the possibility of obtaining a better picture of nature than by way of atomism. Just in order to obtain a standard of comparison for possible new pictures of this kind I will strive in this book to develop the old pictures of mechanics as clearly and consistently as I can. Now let people try to put forward another picture of the world more free from hypotheses, whether on an energetic or purely phenomenological basis, not only just declaring it as possible in a few indefinite hints, but developed from start to finish with the same clarity as we shall now represent the mechanical picture. *Hic Rhodus, hic salta!*

So long as this has not yet been done I admit the possibility but not the certainty that another world picture will displace the mechanical one.

However, pictures that are less definite and distinct than the one to be developed here will be accorded merely a place alongside the latter, since although they are more adaptable, it is more perfect in its inner form, which somehow makes it a paradigm against which every new theoretical idea will have to measure itself as to clarity and definition.

Further, we shall begin from the fundamental assumption of a large though finite number of material points. It is usually said that differential equations avoid a picture that starts from a finite number, but that again is an illusion. Differential equations require just as atomism does an initial idea of a large finite number of numerical values and points in the manifold, that is positions in the manifold of numbers. Only afterwards is it maintained that the picture never represents phenomena exactly but merely approximates to them more and more the greater we choose the number of these points from the manifold and the smaller the distance between them. Yet here again it seems to me that so far we cannot exclude the possibility that for a certain very large number of points the picture will best represent phenomena and that for greater numbers still it becomes less accurate again, so that atoms do exist in large but finite numbers.

The qualitative laws of natural phenomena and their quantitative relations under very simple circumstances, for example the conditions of equilibrium of a heavy parallelipiped of edges in the ratio 1:2:3, can of course be pictured in the mind without starting from a very large finite number of elements. However, as soon as one wants to specify the quantitative laws for complicated conditions one always must start from differential equations, that is first imagine a large finite number of points in the manifold, in short one must think atomistically, and this is not altered by the fact that afterwards we can increase the number of imagined points and so come arbitrarily close to the continuum without ever reaching it.

However this may be, there is a special attraction today in treating mechanics, the most perspicuous scientific discipline, by means of a method that is the very opposite of the modern one and in laying down very special mental pictures from the outset. To begin with the reader may be unable to overcome the feeling that we are merely playing with mental pictures and losing sight of reality. Unperturbed by this we shall first of all try to build up the edifice of ideas as clearly and consistently as possible. If it then agrees with reality, the arbitrary features in the fundamental ideas will thereby have been excused. Indeed we wanted only a picture of nature and by being clearly aware of this, we do not run the danger of trusting the picture more than reality and becoming blind to the latter.

§2. *Fundamental Concepts Borrowed from the Theory of Space and Time.*
First Fundamental Assumption. Continuity of Motion

Every change of place occurs in the course of time and unfolds itself in time. The theory of space as such and time as such must therefore be presupposed, before we can start on mechanics. Time as a manifold of one dimension can be represented by the one dimension of the manifold of ordered numbers, whereas the circumstances of space give rise to a special science, namely geometry. The entire theory of rational and irrational numbers, infinitesimal calculus included, and geometry as well, we therefore presuppose as known. Certainly these sciences have their own difficulties of principle but these are transmitted equally to all views concerning the principles of mechanics since all must start from space and time. Since we here wish to discuss only the difficulties of principle peculiar to mechanics, we shall not bother about those of arithmetic and geometry.

Evidently we cannot obtain a picture of bodies and their motions if we uniformly consider all parts of the whole of infinite space. Let us therefore emphasize a large number of individual points in it as against the rest. These selected points we call material points.

To define the position of any material point at any time we imagine that at all times there is in space a definite rectangular co-ordinate system. By the place of our material point at the given time we understand its position relative to that co-ordinate system which we determine in known ways by Cartesian or polar or some other co-ordinates.

The co-ordinate system is of course nothing real, but this offers no difficulty according to the views we base ourselves on here, since we are at present concerned only with construction and mental pictures. Later we shall see that this co-ordinate system may be chosen in various ways. We shall further see that the place of a material point may be defined instead of by its relative position to our co-ordinate system by that to three specially chosen material points or to a body provided that the material points or body in question have certain properties; or the place may be defined in relation to certain straight lines or planes to be derived from the totality of points, so that we need admit into our mental pictures only one kind of item, namely material points, not a co-ordinate system as well. However, it would merely confuse our picture if we were to take this into account now. Our picture therefore consists of the co-ordinate system and all material points which at all times have a given position relative to the co-ordinate system. That we might choose other co-ordinate systems as well without losing any of the agreement with experience will not disturb us at present.

Just as the question concerning the possibility of determining absolute positions in space offers no difficulty from our point of view, neither does the question concerning a criterion for the equality of various time intervals. We assume that it is possible to construct a perfectly immutable chronometer, to protect it sufficiently from disturbing influences and to replace it by another similar one before it has become in the least worn with use. A glance at the chronometer then tells us about the value of the independent variable we have called time.

I am far from imagining that it is possible exactly to define every word before use either here or later (see the first page of my essay 'On the Question of the Objective Existence of Processes in Inanimate Nature'*, above). The cause why the above pictures are clear is obvious: they are prescriptions for thinking spatial circumstances that everybody can easily and palpably represent for himself in approximation, by means of ruler and pencil or wooden sticks and knitting needles, and which are so well known that their mere idea usually is sufficiently clear even without drawing. A minimum of ideas is employed. The transition from a few individual imaginable points to very many is achieved through general rules. The more we can represent by means of these simple pictures, which we can in any case at present not dispense with in the representation of certain phenomena, the more comprehensible nature must seem to us.

Anything explicable only with the help of further ideas appears much more incomprehensible to us. At all events, when such other ideas are presented one should not just give some indefinite hints concerning the elements from which one starts but honestly make them just as clear and precise as I am trying to do here.

Let us now further develop our picture by assuming certain fictitious laws for the way these material points change place with time. First assumption: we imagine that no two different material points coincide or are infinitely near to each other at any time, but that whenever at any time any material point is at any place (relatively to our co-ordinate system, of course) then also one and only one material point will be at an infinitely near place at any infinitely near time. We say the second material point is the same as the first and call this the law of continuity of motion. It alone enables us to recognize the same material point at different times. The concept of all places at which one and the same material point is in the course of time is called the path of this material point, and the concept of those places that it traverses in a given finite time is called the path during this time.

We may now formulate the law of continuity thus: to every material

---

* *Editor's note*: *Populäre Schriften*, Essay 12, pp. 57–76 in the present collection.

point which at a given time had given co-ordinates, there corresponds at an infinitely near time one and only one material point with co-ordinates differing from the former by infinitely little, and this is called the same material point; that is the co-ordinates of every material point are continuous functions of time $x=\phi(t)$, $y=\chi(t)$, $z=\psi(t)$.

§3. *Second Fundamental Assumption. Existence of Differential Coefficients of the Co-ordinates with Regard to Time. Concept of Velocity and Its Components*

Let us further complete our picture by assuming that the functions $\phi(t)$, $\chi(t)$, $\psi(t)$, expressing the way the co-ordinates of any material point depend on time, have first and second differential coefficients which nowhere become infinite, and this we shall call our second fundamental assumption. Let the co-ordinates of two points A, A' at which a definite material point was at times t, $t+\tau$ be x, y, z and $x_1=x+\xi$, $y_1=y+\eta$, $z_1=z+\zeta$ respectively. Let σ be the length of the line AA'. If now t remains constant and τ becomes ever smaller, the following must result: the quotient ξ/τ approaches a certain limit, which we shall denote by u, or, using differential calculus, by dx/dt or $\phi'(t)$. Similarly for η/τ and ζ/τ. Thus

$$\lim \frac{\xi}{\tau} = u = \frac{dx}{dt} = \phi'(t),$$

$$\lim \frac{\eta}{\tau} = v = \frac{dy}{dt} = \chi'(t), \tag{1}$$

$$\lim \frac{\zeta}{\tau} = w = \frac{dz}{dt} = \psi'(t).$$

These magnitudes may be positive, negative or zero and are called the components of velocity of the material point in the three co-ordinate directions.

Since for every value of τ we have the equation $\sigma=+\sqrt{(\xi^2+\eta^2+\zeta^2)}$, the quotient σ/τ approaches the limit

$$c = \sqrt{u^2 + v^2 + w^2} = \sqrt{\left(\frac{dx}{dt}\right)^2 + \left(\frac{dy}{dt}\right)^2 + \left(\frac{dz}{dt}\right)^2} =$$
$$= \sqrt{[\phi'(t)]^2 + [\chi'(t)]^2 + [\psi'(t)]^2}, \tag{2}$$

which is called the velocity of the material point at time t and can be only positive or zero.

If this last limit is not zero, the direction of the line AA' likewise approaches a definite limiting direction drawn in a definite sense in space, which is called the direction of the material point's velocity at time t. Let us denote in general the angles of an arbitrary straight line G (drawn in a definite sense) with the co-ordinate axes by (G, x), (G, y), (G, z), and the angle between two such lines G and H by (G, H). Since for every value of τ we have $\xi = \sigma \cos(\sigma, x)$ and two similar relations for the other two axes, we further have

$$u = c \cos(c, x), \quad v = c \cos(c, y), \quad w = c \cos(c, z). \tag{3}$$

Let A'', A''', .. $A^{(n)}$ be the spatial points occupied by the material point at times $t+2\tau$, $t+3\tau$ .. $t+n\tau = T$, then, as is shown by integral calculus, the sum of the lines AA', $A'A''$, .. $A^{(n-1)}A^{(n)}$ for diminishing τ and finite constant t and $n\tau$, approaches a definite limit which is called the path traversed in time $T-t$ and in the usual symbolism is denoted by $\int_t^T cdt$, where we often write ds for cdt, with

$$\frac{ds}{dt} = c = \sqrt{\left(\frac{dx}{dt}\right)^2 + \left(\frac{dy}{dt}\right)^2 + \left(\frac{dz}{dt}\right)^2}. \tag{4}$$

The first of the Equations (1) merely means that the increment ξ of the abscissa differs from the increment of time τ multiplied by u by an amount which divided by τ approaches zero for diminishing τ. The meaning of such equations is expressed with special intuitive succinctness if the increments of variables are from the outset denoted by the differential sign put before the variables in question. The equality of two differential expressions then merely means that they differ by a magnitude which, when divided by one of the differentials (if the coefficient of one of them is zero, that one must be chosen), approaches zero (is infinitely small to a higher order) as the denominator decreases. The equality of two differential expressions is therefore established if we can show geometrically or otherwise that their difference is infinitely small to a higher order.

We will often use equations between differential expressions in this sense and therefore write Equations (1), (2), (4) more simply thus:

$$dx = udt, \quad dy = vdt, \quad dz = wdt,$$
$$ds = cdt = \sqrt{(dx)^2 + (dy)^2 + (dz)^2}. \tag{4a}$$

The last of these asserts that the difference between the very small path *ds* and the product of *c* with the time *dt* in which it has been traversed, divided by *dt* approaches the limit zero, whence it follows at once that the sum ∫*ds* of all paths traversed in a finite time equals the integral ∫*cdt* if *c* is given as a function of time.

As is well known, there are functions which for every infinitely small increment of the argument increase infinitely little without the ratio of the increments of the function to that of the argument approaching a definite limit when the latter becomes vanishingly small, and that for any value of the argument; thus the second assumption by no means follows from the first. Anybody who has studied mechanics will remember the difficulty he had in understanding the proof that motion during a very short time can be regarded as uniform and rectilinear and the forces during such a time as constant. These difficulties reside in the fact that these proofs are simply not true.

We have made analytical functions into a representation of the facts of experience. That these functions are differentiable cannot be taken as proof that empirically given functions are equally so, since the number of conceivable undifferentiable functions is just as infinitely great as that of differentiable ones. Likewise the fact that every line drawn by hand or machine corresponds to the aspect of a differentiable function proves merely that as far as our present means of observation go, it is something given in experience that the empirical functions occurring in mechanics are differentiable.

That is why without any further ado we have simply assumed differentiability as being in agreement with the facts of experience to date.

§4. *Introduction of Vectors*

In what follows, vector calculus will often be useful. Let us therefore explain its fundamental concept in the present simple case. By a vector we understand a finite straight line of definite length, direction and sense (indication which of its terminal points is to be the beginning and which the end). Since the purpose of a vector is merely visibly to represent these three things, it is indifferent, so long as it is not yet to express anything else, from what spatial point it is drawn. Most often we are going to draw it from the origin of co-ordinates, *O*, choosing it as the beginning of the line segment in question.

By the sum of two vectors (vector sum) we understand a third vector obtained by placing the second at the end of the first and joining the beginning of the first to the end of the second. The sum is thus obtained like the resultant of two forces in the parallelogram of forces. A vector whose sum with a second yields a third is called the difference (vector difference) of the third and second. If the projections of a vector on the x-axis equals the sum of such projections of two other vectors and likewise for the other two axes, then, as is easily seen, the first vector is the sum of the other two. Likewise for the difference of the two vectors and the sum of more than two. To find the latter one must add the third to the sum of the first two and so on: from the end point of the first vector AB draw a line BC equal in length and direction to the second vector, from the end point C of this line a line CD equal in length and direction to the third vector and so on. The line from the beginning A of the first vector to the end point M of the last one is then the sum of them all. The figure

$$ABCD \ .. \ M \tag{5}$$

which, of course, need not lie in a plane, is called the vector polygon; in the special case where the vectors represent forces, it is the polygon of forces.

Evidently the position of arbitrary material points at any time may be represented by the vectors drawn from an arbitrary spatial point, for example the origin of co-ordinates, to the spatial points where the material points are at the time in question. The distance AA' between the spatial points A and A' where a material point is at times t and $t+\tau$ can then be regarded as a vector too, namely the difference between the two vectors OA' and OA that join the points A and A' to the origin O. From an arbitrary point, for example the origin, we can draw a vector Ob equal in length and direction AA'. Since, however, the length of AA' becomes infinitesimally small with vanishing τ, we can increase the length of Ob in the ratio that some arbitrary fixed time chosen once and for all (the unit of time) bears to τ. The limit OB which the vector thus magnified approaches with vanishing τ is called the velocity vector of the material point in question at time t. It represents the velocity of the material point in magnitude and direction, and its projections on the three co-ordinate axes the components u, v, w of the velocity, irrespective of the point from which the vector is drawn.

§5. *The Concept of Acceleration and Its Components*

The second assumption further contains the condition that the co-ordinates possess second differential coefficients with respect to time. If, during a finite time $n\tau$ these latter are zero, the joins AA', $A'A''$, $A''A'''$ .. of the spatial points occupied by the material point at times t, $t+\tau$, $t+2\tau$, .. $t+n\tau$ all lie on the same straight line and are equal in length: the path of the material point is thus a straight line and equal paths are traversed in equal times. The velocity is constant, the motion is uniform. Every acceleration or retardation of motion, every curvature of the path to one side or the other is thus conditioned by the second differential coefficients of co-ordinates with respect to time being different from zero. It is above all their values that must now be studied.

As before, let a material point at times t, $t+\tau$, $t+2\tau$, occupy spatial points A, A', A'' with projections D, D', D'' on the x-axis (see Figure 1).

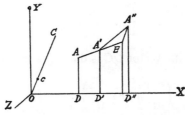

Fig. 1.

Let the co-ordinates of the three points be x, y, z; x', y', z'; x'', y'', z''. As is well known, the second differential coefficient of the material point's abscissa x with respect to time, which is called the component of accelera-tion of the material point in that direction, equals the limit of the expression

$$\frac{x'' - x' - (x' - x)}{\tau^2} = \frac{D'D'' - DD'}{\tau^1} \tag{6}$$

for vanishing τ; but the segments DD' and $D'D''$ are the projections of the vectors AA' and $A'A''$ on the x-axis. The numerator of the fraction in (6) is thus the projection of the difference between these two vectors on the x-axis. Since it is obviously indifferent from what point the vectors are drawn, let us choose the same spatial point for both vectors, for example replacing AA' by another vector $A'E$ equal in length and direc-tion to the former but drawn from A'. The join EA'' of the terminal points of the two vectors $A'E$ and $A'A''$ or a line Oc parallel to it and and equal in length but drawn from the origin of co-ordinates thus represents the difference of the two vectors $A'A''$ and AA'. The projection of Oc or EA'' on the x-axis is thus equal to the numerator of Formula (6). The limit that this difference divided by τ^2 approaches is the component of the acceleration in the x-direction. Since the same holds for the other two directions, the vector EA'' or Oc gives us an accurate picture of the curvature of the path and of the acceleration or retardation of the motion. It will be preferable to draw the length magnified in the ratio of the time unit squared to τ^2. The limit OC which the vector so magnified drawn from the origin at constant t approaches for vanishing τ is called the acceleration vector or more briefly the acceleration of the material point in question at time t. Its component in the three co-ordinate directions are equal to what we have already called the components of acceleration in the three co-ordinate directions, that is to

$$\frac{d^2x}{dt^2}, \quad \frac{d^2y}{dt^2}, \quad \frac{d^2z}{dt^2}.$$

The total length of the vector OC is the positive square root of

$$\left(\frac{d^2x}{dt^2}\right)^2 + \left(\frac{d^2y}{dt^2}\right)^2 + \left(\frac{d^2z}{dt^2}\right)^2.$$

Of course one would have obtained the same vector by joining the terminal points of the two velocity vectors OB and OB' of the material point at times t and $t+\tau$ and seeking the limit that this line divided by τ approaches.

Let (g, x), (g, y), (g, z) be the angles between the acceleration of our material point (that is of the vector OC) and the three co-ordinate axes and g the magnitude of this acceleration (that is the length of this vector):

$$\frac{d^2x}{dt^2} = g\cos(g,x), \quad \frac{d^2y}{dt^2} = g\cos(g,y), \quad \frac{d^2z}{dt^2} = g\cos(g,z). \quad (7)$$

When the acceleration of a material point is represented in one case by the vector OC and in another by OD, we understand by the sum of these two accelerations that which is represented by the sum of the two vectors OC and OD. If the components of the acceleration OC in the co-ordinate directions are

$$\frac{d^2x_1}{dt^2}, \quad \frac{d^2y_1}{dt^2}, \quad \frac{d^2z_1}{dt^2}$$

and those of OD

$$\frac{d^2x_2}{dt^2}, \quad \frac{d^2y_2}{dt^2}, \quad \frac{d^2z_2}{dt^2},$$

then the components of the sum of the two accelerations are

$$\frac{d^2x_1}{dt^2} + \frac{d^2x_2}{dt^2}, \quad \frac{d^2y_1}{dt^2} + \frac{d^2y_2}{dt^2}, \quad \frac{d^2z_1}{dt^2} + \frac{d^2z_2}{dt^2}.$$

Similarly we define the sum of three or more accelerations.

§6. *Fundamental Assumptions 3–7*

Let any number (n) of material points be given. At any time t let them be at spatial points $A_1, A_2, A_3, .. A_n$. To complete the picture we make two further fundamental assumptions that the accelerations will help us to find from the constellation of the material points.

Third fundamental assumption: the acceleration of any material point, as a vector in the above sense, is equal to the sum of $n-1$ accelerations, in the same vectorial sense, of which each has the direction of the join between this and one of the other material points and is called the acceleration the second point imparts to the first or the acceleration of the first produced by the second.

Fourth fundamental assumption: the acceleration of any material point by another is always in the opposed direction to that of the second by the first. If the first acceleration has the direction of the join from the first to the second material point (Case B) the second acceleration lies along the same join in the opposite direction, and the two material points attract each other. If, however, the first acceleration lies in the prolongation of the join away from the second point (Case A), then the second acceleration too lies in the other prolongation away from the first point, and the two material points repel each other.

Fifth fundamental assumption: the magnitude of the acceleration g_{12} of an arbitrary material point by any other depends neither on their absolute position in space nor on the absolute value of time, nor on the character of the surroundings or the velocity of the point in question, nor on the direction of their join in space, but only on the latter's length r_{12}. It is thus a function $F(r_{12})$ dependent only on that length. We set $F(r_{12})$ positive or negative, putting $g_{12} = +F(r_{12})$ or $g_{12} = -F(r_{12})$ (since g_{12} as a vector is always taken as positive) according to whether the two points repel or attract each other, that is whether the acceleration falls into the prolongation of the join or into the join itself. The form of this function will be left entirely indefinite for the time being.[3]

Sixth fundamental assumption: the magnitude of the acceleration imparted by the first material point to the second need not be the same that the second imparts to the first, but both stand in a permanently constant ratio at all distances. Putting the magnitude g_{21} of the acceleration of the second material point by the first equal to $\mu_2 F(r_{12})$, then μ_2 is a constant magnitude for this point pair at all times and distances. It is essentially positive since for the second point we put $g_{21} = +\mu_2 F(r_{12})$ for attraction and $g_{21} = -\mu_2 F(r_{12})$ for repulsion.

Seventh fundamental assumption: if r_{13} is the distance of our first material point from an arbitrary third material point and $\Phi(r_{13})$ the acceleration of the first material point by the third, and $\mu_3 \Phi(r_{13})$ that of the third by the first, then the acceleration of the second material point by the third and that of the third by the second stand in the permanently constant ratio of $\mu_2 : \mu_3$; so that if r_{23} is the distance between them and $\Psi(r_{23})$ the acceleration of the second point by the third, then $(\mu_3/\mu_2) \times \Psi(r_{23})$ must be that of the third by the second.

To express this assumption more symmetrically, we denote by m_1 some quite arbitrary though permanently and everywhere constant positive number and put $m_1/\mu_2 = m_2$; $m_1/\mu_3 = m_3$.

Moreover we denote the magnitude $m_1 \cdot F(r_{12})$, which evidently must also be a function of r_{12} with the same sign as $F(r_{12})$, by $f_{12}(r_{12})$; and the magnitudes $(m_1/\mu_3) \, \Phi(r_{13})$ and $(m_1/\mu_2) \, \Psi(r_{23})$ by $f_{13}(r_{13})$ and $f_{23}(r_{23})$. We can then write the above relations in the following symmetrical form

$$m_1 g_{12} = m_2 g_{21} = \pm f_{12}(r_{12}),$$
$$m_1 g_{13} = m_3 g_{31} = \pm f_{13}(r_{13}),$$
$$m_2 g_{23} = m_3 g_{32} = \pm f_{23}(r_{23}),$$

where in the case of repulsion we use the positive sign and for attraction the negative sign.

If we consider a fourth material point as well, experience indicates the accelerations $g_{14}=F_1(r_{14})$, $g_{24}=\Phi_1(r_{24})$, $g_{34}=\Psi_1(r_{34})$ imparted to the first, second and third points by the fourth at distance r_{14}, r_{24}, r_{34}. Moreover, we observe the factor μ_4 with which we must multiply the acceleration $F_1(r_{14})$ to obtain the acceleration g_{41} of the fourth material point by the first. Since the seventh fundamental assumption is to hold for any three arbitrary material points, the acceleration g_{42} of the fourth material point by the second must be equal to $(\mu_4/\mu_2)\,\Phi_1(r_{24})$ and g_{43} of the fourth by the third to $\mu_4/\mu_3\Psi_1(r_{34})$. Putting exactly as before

$$\frac{m_1}{\mu_4} = m_4, \quad m_1 F(r_{14}) = f_{14}(r_{14}),$$

$$\frac{m_1}{\mu_2}\, \Phi_1(r_{24}) = m_2\Phi_1(r_{24}) = f_{24}(r_{24}),$$

$$\frac{m_1}{\mu_3}\, \Psi_1(r_{34}) = m_3\Psi_1(r_{34}) = f_{34}(r_{34}),$$

we obtain

$$m_1 g_{14} = m_4 g_{41} = \pm f_{14}(r_{14}),$$

$$m_2 g_{24} = m_4 g_{42} = \pm f_{24}(r_{24}),$$

$$m_3 g_{34} = m_4 g_{43} = \pm f_{34}(r_{34}).$$

Extending this to more than four material points offers no difficulty.

§7. *Mass and Force. Equality of Action and Reaction*

According to what has been said in the previous paragraph we obtain for any material point a definite number m, which we call its mass, and for any two material points a function $f(r)$ of their distance r, which we call the force acting between these two points at that distance. The absolute value of $f(r)$ is called the intensity of the force, whether of the first point on the second which equals the product of the mass of the first and its acceleration by the second, or of the second on the first, which equals the product of the mass of the second and its acceleration by the first. The intensity of the force is to be regarded as essentially positive, like the absolute value of the acceleration. The direction of the acceleration imparted by the second point to the first is called the direction of the force exerted by the second on the first. It is always opposite to the direction of the first exerted on the second and acting towards or away from the second according to whether $f(r)$ is negative or positive.

The force of a material point exerts on a second is thus always equal but opposite in direction to the force the second exerts on the first. One also says that action and reaction are equal but opposite in direction.

For brevity's sake we say that the force acting between the two points is a central force by which we express that its intensity is a function only of their distance, that its direction is that of their join and that action and reaction are equal and opposite in direction. In order that the motion is certainly and unambiguously determined we further assume that the necessarily unambiguous function of the distance r that gives the force has, for all relevant values of r, a finite first derivative (including zero) or at least that the ratio of the increments of $f(r)$ and r never becomes infinite for these values of r.

Of the masses m of all material points one is quite arbitrary. All the others are then determined by this one and by facts of experience.

Although, rather than give references, I prefer to start with the remark that in this book I present only what is well known and lay no claim to having found any of the theorems quoted, nevertheless I must here mention that the above definition of mass is due to Mach,[4] since this fact is perhaps less well known.

It would simplify our picture if we took all material points to have equal masses, that is, if we assumed that any two material points impart equal but opposite accelerations to each other. If we then assumed that in denser bodies there are simply more material points for each unit of volume, we could represent all phenomena just as adequately. A material point of mass m could equally be represented fairly approximately as m very close and rigidly connected material points of mass 1. I have adopted the more general picture only because it too is quite clear and definite.

Of course, the fundamental assumptions here made are organically interconnected, so that we should fall into frequent contradictions if we dropped or altered only one or the other while keeping the rest unchanged. For example, it can be shown that if we assume that the acceleration that two material points impart to each other does not depend on their position in space and on the absolute value of time, but drop one of the assumptions 4, 5 or 7, the velocity of two rigidly connected material points could in time become infinite, where of course it is further presupposed that the action of the device that rigidly and closely connects them is itself produced by forces that obey our assumptions. From this it obviously does not follow that the totality of our assumptions could be replaced by the principle of energy or by some other general principles. The possibility that some of our assumptions might be thus replaced I will not deny.

Indeed, instead of starting from the concept of acceleration one might start from the equation of kinetic energy, for example by presupposing that this equation holds separately for each co-ordinate direction. Given the important role of the principle of energy throughout nature this approach might well appeal to some. However, I have found it impossible to replace the fundamental assumptions here made by more general principles in such a way that the fundamental assumptions really become significantly simplified. Therefore I have made no special efforts, since it seems to me not at all essential or promising once one has decided in any case to start from action at a distance of material points and only later to deduce Hamilton's principle, the equations of elasticity and hydrodynamics and so on.

§8. *General Equations of Motion*

If the force between two material points with co-ordinates x_1, y_1, z_1 and x_2, y_2, z_2 respectively is repulsive, the acceleration g_{12} of the first by the second falls, as mentioned, into the prolongation of their join r_{12} from 2 to 1, which with the positive co-ordinate axes makes angles whose cosines are

$$\frac{x_1 - x_2}{r_{12}}, \quad \frac{y_1 - y_2}{r_{12}}, \quad \frac{z_1 - z_2}{r_{12}}.$$

These are the cosines of the angles (g_{12}, x), (g_{12}, y), g_{12}, z) of the acceleration g_{12} with the positive co-ordinate axes. In that case we give the function $f_{12}(r_{12})$ a positive sign too. If then the first material point were accelerated only by the second, Formula (7) would give

$$\frac{d^2 x_1}{dt^2} = g_{12} \cos(g_{12}, x) = \frac{1}{m_1} f_{12}(r_{12}) \frac{x_1 - x_2}{r_{12}},$$

with two analogous equations for the other two axes. If the acceleration is in the opposite direction, we take it again as positive but give the opposite sign to the cosines, so that

$$\cos(g_{12}, x) = \frac{x_2 - x_1}{r_{12}}.$$

But since the function would then also be given the opposite sign, we still have the above equation.

According to the third fundamental assumption the total acceleration of the first material point is the vector sum of the different accelerations imparted to it by the other material points, and according to (8) the total component of acceleration in the x-direction is then the ordinary algebraic sum of the individual components of acceleration. Thus we have in general

$$m_1 \frac{d^2 x_1}{dt^2} = \sum_{k=2}^{k=n} f_{1k}(r_{1k}) \frac{x_1 - x_k}{r_{1k}} \tag{9}$$

and two analogous equations for the other two co-ordinates and $3n-3$ similar equations for the other material points.

The values of the magnitude over which we are to sum, indicated above and below the summation sign, express (as always in what follows) that in the expression behind the sign this magnitude is to run through all the values from the lower to the upper inclusive and all such terms are to be added together. We denote by $\phi_{hk}(r_{hk})$ the indefinite integral $\int f_{hk}(r_{hk}) \, dr_{hk}$ in which the constant of integration may be given any special value.

Further we shall often have to give an increment to a single co-ordinate in an expression containing the co-ordinates of all the material points in a given context, while all the other co-ordinates in the expression remain constant. Increments occurring in this manner are called partial and are denoted by the symbol ∂. They must not be confused with increments occurring during the time dt (total increments); for during this time generally all co-ordinates change. Thus the partial differential coefficient $(\partial r_{hk}/\partial x_h)$ of r_{hk} with respect to x_h has the following meaning: of all the co-ordinates only x_h is given a small increment, the corresponding increment of r_{hk} is divided by it, and then we take the limit this expression approaches for vanishing increment of x_h. Since

$$r_{hk} = \sqrt{(x_h - x_k)^2 + (y_h - y_k)^2 + (z_h - z_k)^2}$$

differential calculus gives

$$\frac{\partial r_{hk}}{\partial x_h} = \frac{x_h - x_k}{r_{hk}}, \quad \frac{\partial r_{hk}}{\partial y_h} = \frac{y_h - y_k}{r_{hk}}, \quad \frac{\partial r_{hk}}{\partial z_h} = \frac{z_h - z_k}{r_{hk}}.$$

Moreover from the definition of the functions ϕ it follows that

$$\frac{\partial \phi_{hk}(r_{hk})}{\partial x_h} = \phi'_{hk}(r_{hk}) \frac{x_h - x_k}{r_{hk}} = f_{hk}(r_{hk}) \frac{x_h - x_k}{r_{hk}}$$

and similarly for y and z.

In the expression $\phi_{hk}(r_{hk})$, let h run through the integers from 1 to n inclusive, and for every h let k similarly run through the same integers leaving out h. The sum of all expression thus obtained is denoted by

$$\sum \sum \phi_{hk}(r_{hk}) = - V. \tag{10}$$

Equation (9) can then be written in the form

$$m_1 \frac{d^2 x_1}{dt^2} = - \frac{\partial V}{\partial x_1}, \tag{11}$$

with two similar equations for the other two co-ordinate axes and $3n - 3$ more for the other material points. The force component acting on any material point in any co-ordinate direction is thus the negative partial derivative of the function V (the force function) of all co-ordinates with respect to the co-ordinate in question.

For my feeling there is still a certain lack of clarity in the differential coefficients with respect to time. Except for the few cases where one can find an analytic function that has exactly the prescribed differential coefficients with respect to time, then in order to set up a numerical picture one will always have to imagine time as divided into a finite number of parts before one proceeds to the limit.[5] Perhaps our formulae are only very closely approximate expressions for average values that can be constructed from much finer elements and are not strictly speaking differentiable. As to that, however, there are so far no indications from experience.

§9. Different Modes of Expression. Resultants. Components

Instead of saying that the second material point imparts the acceleration g_{12} to the first one we can also say that the force $\pm f_{12}(r_{12})$ that the second exerts on the first produces this acceleration; but we must not forget that these are merely different words for one and the same fact. As mentioned earlier, the absolute value of the function $f_{12}(r_{12})$ which is the product of the mass m of the material point on which it acts and the acceleration g that it imparts to that point, we denote as the intensity of the force and the direction of the acceleration as the direction of the force. Just as with accelerations, therefore, we can express forces as vectors (arrows) whose length equals the intensity of the force in question and whose direction coincides with that of the force and therefore also with the acceleration produced by it. These vectors representing forces are usually drawn not at the origin but from the point of application of the force.

The vector sum of all forces acting on a material point again represents a force which is called their resultant; the individual forces are called its components. Since according to our third fundamental assumption the actual acceleration of a material point is the vector sum of the different accelerations it would acquire by each individual force, so that the force vectors differ from the acceleration vectors only in that the former are m times longer, it follows that the actual acceleration of a material point has the direction of the resultant force, and its product with the mass of the material point equals the intensity of the resultant force. Thus we find the actual acceleration from the resultant force in the same way as the acceleration caused by a single force from it.

We can easily generalize this further still. Let the vector R be the sum of any other vectors C, then the acceleration imparted to a material point by a force represented by R is the vector sum of the accelerations that the various forces represented by the vectors C would impart to the point. Since it is indifferent in which order vectors are added into a sum and since force vectors are always m times longer than the corresponding acceleration vectors but parallel to them and since according to the third fundamental assumption accelerations are added like vectors, the material point will be given the same acceleration by the resultant force R as by all the component forces C, whether or not there are any other forces also acting on it, contributing their own additional accelerations.

The sum of two vectors can be a zero vector only if both have the same length but are opposite in direction. A material point on which two forces are acting will thus undergo no acceleration, that is behave as though no force were acting on it, if and only if those two forces have equal intensity but opposite direction. We then say that the forces are in equilibrium. Since the resultant of any number of forces acting on a material point is found by means of the figure (5) under §4 above, (the polygon of forces, for two forces a parallelogram), the forces will be in equilibrium, that is not impart any acceleration to the point, if the polygon of forces $ABC .. M$ is closed so that its end point M coincides with its initial point A.

From the construction of the polygon of forces we see that the action of a force P can be completely replaced by three forces acting along the co-ordinate directions, namely

$$X = P \cos(P, x), \qquad Y = P \cos(P, y),$$
$$Z = P \cos(P, z) \tag{12}$$

which are called its components in the co-ordinate directions.

We denote by P_1 the resultant of all forces acting on the material point of mass M_1 to which Equations (9) refer, with $P_1', P_1'' ..$ any components into which the force P_1 can be decomposed, with X_1, Y_1, Z_1; $X_1', Y_1', Z_1'; X_1'', Y_1'', Z_1'' ..$ the components of $P_1, P_1, P_1'' ..$ in the co-ordinate directions. Then we can rewrite (9) thus:

$$m_1 \frac{d^2 x_1}{dt^2} = X_1 = X_1' + X_1'' + . \tag{13}$$

Similar equations hold of course for the other co-ordinate axes and material points.

§10. *Poisson's Proof of the Parallelogram of Forces*

Of the many proofs that have been given of the theorem of the parallelogram of forces let me here briefly explain in somewhat modified form, though only as an ideal picture, the proof provided by Poisson in his mechanics.

Let us assume that we can always replace several forces (the components) acting on a point by a single force (the resultant) as regards all
their effects. From this it follows that we can always find a resultant of
more than two forces by first combining two into their resultant, then
combining that with a third into a new resultant and so on. For since the
first resultant completely represents the first two forces, the new resultant
which always has the same effect as the first resultant and the third force
must likewise have the same effect as the first three forces and so on.

We further presuppose that the intensity of the resultant and its
position relative to the components is independent of the absolute
spatial position of the figure, of the circumstances of motion, earlier
temporal conditions and origin of the forces, so that forces of any
origin behave in the same way. A force that produces the same effect as
two perfectly equal and parallel forces together we call a force of twice
their intensity; a force of three times the intensity is one whose effect
equals that of three equal but parallel forces together and so on. As to
the laws or the import of the motions produced, we are not here concerned.

If two perfectly equal forces act on a point but in opposite directions,
this determines no motion either collinear with them or at right angles.
The point, if originally at rest, must therefore continue to be so and
equilibrium must prevail; the point must behave as though no forces were
acting on it.

If therefore a point is acted on by a force in one direction and a force
of twice the intensity in the opposite direction, we can replace the latter
by two forces of single intensity of which one is balanced by the force in
the opposite direction. Thus only one is left.

Continuing the same line of inference we easily see that the resultant of
any number of forces whose direction is in a given line is their algebraic
sum, counting forces as positive in one sense and negative in the opposite
sense; the resultant, too acts in one sense or the other according to the
sign of the algebraic sum being positive or negative.

Let a point A be acted on by two equal forces AB and AC with any
angle between them. Let us call the part AH of the angular bisector
between them the positive sense, the other part thus being negative.
Let α be the angle between one of the forces and the positive bisector so
that the angle between the two forces is 2α. The angle may for the present
be acute or obtuse, zero or equal to one or two right angles.

The only straight line determined unambiguously by two straight lines from a point is their angular bisector. The resultant AD of the forces AB and AC must therefore fall on to this bisector, whether on the positive or negative side.

If without changing the direction of the two forces we double their intensities, we can view the matter as though in the original direction of AB there were now two equal forces AB and AB_1 and similarly AC and AC_1 for the other. The resultant of AB and AC is AD, that of AB_1 and AC_1 is AD_1. AD and AD_1 together make a resultant of equal direction but twice as big as AD. In this way one proves that AD must be proportional to the intensity of the equal forces AB and AC. However, it might still depend on the angle between them, so that we may put $AD = AB\,f(\alpha)$, where the function $f(\alpha)$ is given the positive sign when AD falls on the positive part AH of the bisector, and the negative sign if AD falls on the negative part.

If we put $180 - \alpha$ for α, the two forces make the same angle in the opposite direction, so that the resultant too must simply be reversed in sign, and $f(180 - \alpha) = -f(\alpha)$, so that we do not restrict generality by presupposing in what immediately follows that α does not exceed $90°$. Let us now draw a straight line AK making with AH any angle β lying between α and $90°$ towards the side on which the force AB is situated.

Moreover let the point A be acted on by two further forces AB' and AC' represented by arrows that are the exact mirror images of AB and AC with respect to AK. Let AH' be the mirror image of AH with respect to AK. The two forces AB' and AC' then have a resultant AD' which likewise is the mirror image of AD with respect to AK, so that the resultant of all four forces AB, AC, AB', AC' is obtained by finding that of AD and AD'. These last two forces are equal and form an angle β with AK or its extension, according as $f(\alpha)$ is positive or negative. The resultant is therefore

$$ADf(\beta) = ABf(\alpha)f(\beta) \tag{13a}$$

acting in the direction AK or in the opposite sense, according as the expression is positive or negative. The same force must also result if we first form the resultant of AB and AB', then that of AC and AC', finally combining those two. The resultant of AB and AB' equals $ABf(\beta - \alpha)$, that of AC and AC', $ABf(\beta+\alpha)$. Both fall into the line AK. Each has the direction AK or the opposite one, according as the expression for them is positive or negative. The resultant of these two resultants is therefore their algebraic sum $AB[f(B-\alpha)+f(\beta+\alpha)]$ acting also in the direction AK or in the opposite one, according as the expression is positive or negative.

Since the resultant of the four forces AB, AB', AC, AC' equals this expression as well as that in (13a), these two must be equal. The sign, too, has the same meaning in both: positive if the action is in the direction AK, negative if in the opposite one. Therefore, for all α and β within the limits assumed for them,

$$f(\alpha)f(\beta) = f(\beta-\alpha) + f(\beta+\alpha).\tag{13b}$$

For $\alpha=0$ the resultant is twice the size of each component. Hence $f(0)=2$. If ε is a very small angle, we can therefore put $f(\varepsilon)=e^{\zeta}+e^{-\zeta}$ or $f(\varepsilon)=-(e^{\zeta}+e^{-\zeta})$ or $f(\varepsilon)=2\cos\zeta$ according as $f(t)>2$ or <-2 or between $+2$ and -2. If it were exactly $+2$ or -2 we should choose the first or second form with $\zeta=0$. Let us now make the following sequence of substitutions in (13b): $\alpha=\varepsilon$, $\beta=\varepsilon$; $\alpha=\varepsilon$, $\beta=2\varepsilon$; $\alpha=\varepsilon$, $\beta=3\varepsilon$ or $\alpha=\beta=2\varepsilon$; $\alpha=\varepsilon$, $\beta=4\varepsilon$ or $\alpha=2\varepsilon$, $\beta=3\varepsilon$ and so on; integral h gives $f(h\varepsilon)=e^{h\zeta}+e^{-h\zeta}$ or $=(-1)^h(e^{h\zeta}+e^{-h\zeta})$ or $=2\cos(h\zeta)$, according as the first, second or third form of $f(\varepsilon)$ has been chosen. The resultant vanishes for $h\varepsilon=90°$, that is $f(90°)=0$, but not for any value of $h\varepsilon$ between 0 and $90°$; hence $f(h\varepsilon)$ cannot be represented by one of the exponential formulae. Therefore it must equal $2\cos(h\zeta)$ and for $h\varepsilon=90°$ we must have $h\zeta=90°$, so that $\zeta=\varepsilon$. Therefore $f(h\varepsilon)=2\cos(h\varepsilon)$ and replacing any value of $h\varepsilon$ by the symbol α, we have $f(\alpha)=2\cos\alpha$, which proves the parallelogram of forces for the case of two equal components.

Next we proceed to the case of two unequal forces AB and AC at right angles acting on a point A. Bisect the line BC at D and decompose each of the given forces into one component along AD and another that makes the same angle with the force being decomposed but on the opposite side. The latter two components cancel each other, the two former are each equal to AD and together yield the resultant of the two original components as given by the parallelogram of forces.

Only at this stage can we determine the resultant of any two forces AB and AC acting on a point A. We draw the parallelogram $ABDC$, decompose each force into two components of which one lies along AD and the other at right angles to it. We see at once that the latter two components cancel and the first two again yield the resultant represented by the diagonal AD of the parallelogram.

Of course this is by no means a proof that all our previously made assumptions are correct. It shows only that we should become entangled in contradictions if for defining force we were to retain the other assumptions but were to make some different assumptions as to constructing the resultant of two forces.

Even the assumption that the intensity of the resultant and its relative position to the components does not depend on the position of the figure in space, or with regard to the fixed stars, is not so self-evident as people imagine, since for example the forces that can permanently maintain a certain system in a certain relative configuration of its parts by no means depend only on this relative position but alter if the whole system rotates in space without change of the relative position of its parts.

§11. *On the Replacement of the Picture's Co-ordinate System by Others*

Since Equations (9) merely determine the second differential coefficients of the co-ordinates with respect to time, we must further be given the $6n$ values of all co-ordinates and of their first differential coefficients with respect to time (the components of velocity), at some time (the initial state). This time is often denoted by t_0 or zero and the values of the co-ordinates and components of velocity at this time by $x_1^0, y_1^0, .. w_n^0$. These $6n$ values and the differential Equations (9) then unambiguously determine the values of all co-ordinates and velocity components at any other time.

Since in our picture we have based ourselves on one definite co-ordinate system, we must first examine how far the same rules for determining the values of co-ordinates and velocity components from initial values remain valid for other co-ordinate systems too. All co-ordinate systems for which this is so, together with the one orininally chosen, we call suitable reference frames. Let us first imagine that any other co-ordinate system has been introduced, with axes always parallel to those of the original ones and remaining in the same relative position to them. Then the co-ordinates of any point relative to this new system differ from those relative to the original one only by additive constants; while the velocity

components, accelerations and all terms on the right-hand side of the
Equations (9), which evidently depend only on the relative position of
the material points, remain completely unchanged. Thus these equations
remain valid as they stand if in them we replace the co-ordinates relative
to the original system by those relative to the new one. The new co-
ordinate system therefore achieves exactly the same results as the original
one, since the changes of co-ordinates in the new system can be computed
from their initial values and the initial velocity components by exactly
the same rules as were established for the original system. This holds
even if the new axes remain parallel to the original ones while the new
origin of co-ordinates moves at constant speed in a straight line relatively
to the old axes, with component speeds a, b, c, along these latter. In
that case transition to the new co-ordinate system adds the constant a
to all x-components of velocity, and b and c for the other two components;
while the x-co-ordinate of all points is increased by $at + \alpha$, and the other
two by $bt + \beta$, $ct + \gamma$, where α, β, γ are three new constants. Again, this
does not alter the accelerations or the right hand terms of the Equations
(9) and the rules for finding the motions of the system remain valid for
the new system of co-ordinates as much as for the old.

The same is still true if at relative rest or uniform rectilinear motion of
a new rectangular co-ordinate system the latter's axes are not parallel to
those of the original one, although the angles between them remain-
constant. For both the accelerations and the forces have been determined
by the construction of vectors that are independent of the position of
the co-ordinate system; but the expressions for the projections of accelerations and forces in the co-ordinate directions are the same for all co-
ordinate systems. Let the co-ordinates in the new system be marked by
dashes, and the angles between old and new axes by (x, x'), $(x, y') \ldots$
Then

$$\frac{d^2 x_1'}{dt^2} = \frac{d^2 x_1}{dt^2} \cos(x, x') + \frac{d^2 y_1}{dt^2} \cos(y, x') + \frac{dz_1^2}{dt^2} \cos(z, x').$$

If in this we substitute for

$$\frac{d^2 x_1}{dt^2}, \quad \frac{d^2 y_1}{dt^2}, \quad \frac{d^2 z_1}{dt^2},$$

their values from the Equations (9), in which of course we must substitute for x_1-x_2, y_1-y_2, z_1-z_2 the values $(x_1'-x_2')\cos(x_1 x')+(y_1'-y_2')$ $\times\cos(x, y')+(z_1'-z_2')\cos(x, z')$ and so on, and proceed similarly with $d^2 y_1'/dt^2$, $d^2 z_1'/dt^2$, we can easily reduce the equations for the new co-ordinates into exactly the form of the Equations (9).

There are thus very different co-ordinate systems that might be used as the basis for our picture just as well as the original system, without any change to the rules for deriving the motion of the system from the initial values of co-ordinates and velocity components; all of them are suitable reference frames. This is very important, since often the choice of this or that co-ordinate system offers certain advantages. However, the new system must not rotate relatively to the old one, or the new origin move non-uniformly or on curvilinear paths relatively to the old one, if these rules are to be exempt from any change; for in the former case the angles (x, x') and so on would no longer be constant, nor in the latter case the magnitudes denoted by a, b, c, α, β, γ, so that the second differential coefficients of the co-ordinates with respect to time would no longer assume the above form, as will be shown in more detail in Part II. In the latter case the only change would be that all second differential coefficients of x-co-ordinates with respect to time had the same function of time added to it, and similarly for the other two co-ordinates.

§12. *Relation of This Representation to Others*

We have deliberately gone rather far away from reality, in order to obtain as precise and clear a picture as possible, that is, one free from vague concepts but offering the most definite indications for the purpose of calculation, so that in every definite case the result to be expected can be unambiguously and securely predetermined to any degree of approximation.[6] The requirement that the picture should be thus unambiguous seems to me to be what Hertz understands by the requirement that the picture should coincide with the laws of thought; for I cannot really imagine any other law of thought than that our pictures should be clearly and unambiguously imaginable and that from them results always agreeing with experience can be derived as readily as possible. Nor am I in the least of the opinion that anything, say geometrical images, can be derived from the laws of thought alone.

On the other hand, Hertz seems justified when he remarks that most representations of the fundamental principles of mechanics lack the desirable consistency and precision, which I aimed at by fearlessly starting with quite definite hypotheses worked out in detail. One often defines the ratio of masses of a body as the inverse ratio of the accelerations they assume under the action of equal forces. Equal forces can be made to act on extended rigid bodies by putting them on a perfectly smooth horizontal table and then fastening the same equally stretched elastic string or the same equally influenced small magnet or electrified object, now to one body, now to the other. Fluids would have to be enclosed in a container of small mass, to which the above devices can be fastened. As we shall see later from the picture that we can form of the action of neighbouring volume elements of elastic and fluid bodies, we can indeed consider the centre of gravity of the system, body and magnet or container, fluid and magnet, as a single mass on which a force is acting that depends only on the constitution of the magnet and the influences on it. The same holds for the case of a stretched string if its mass vanishes. But, how are we to attach equal forces to individual material points? However, if the definition of the mass ratio of two bodies is derived from their impact, we can in any case not dispense with the consideration of the volume elements that collide; if it is derived from direct action at a distance between two small bodies, then if the derivation is carried through consistently our own definition again results.

If one were to start with a perfectly clear picture of the mutual action of volume elements of elastic bodies and deduced from them the fundamental concepts and laws of ordinary mechanics, there would of course be no objection.[7] However, this is not at all what happens, rather one defines the properties (masses) and laws of change (forces) of simple material points by means of processes that essentially contain such volume elements, for example impact or fastening the same elastic string to different bodies. In the ideas one forms of the material points experience with extended objects are mixed up with conceptual constructions on individual points. Who would not feel the vicious circle involved in defining a material point, for the purpose of setting up the fundamental concepts, as a very small body and then insisting that a remote consequence of the theory built on this basis will be that if we regard volume elements as simple spatial points although they are really small bodies, we are neglecting only small quantities of higher order. How much clearer to view extended bodies from the outset under the picture of many tightly packed points whose velocity always changes only very little from one to the next. Starting from the accelerations of finite bodies without first explaining the concept of the material point is again inadmissible, if we do not have prior knowledge of the theorem concerning the centre of gravity. It is no use objecting that these are merely different words for the same thing. Therefore I think one must choose precisely those words that will always remind us in the most appropriate way of the correct epistemological status of all concepts.

Yet even admitting that application of the same force to different material points without prior definition of mass were possible, we must in addition recognize as facts of experience that the ratio of accelerations of material points does not change according to the choice of force and that the ratio of accelerations of the material points A and C equals the product of the respective ratios for A and B, B and C; or else more general facts of experience (for example the principle of energy) must be advanced first from which those others follow. The advantage of our picture of central forces is precisely that at a stroke it provides a clear view about all these propositions as well as about the laws of mutual action of volume elements, the theorem on centres of gravity and so on.

In Kirchhoff's lectures on mechanics I am also dissatisfied with the definition of the concept of mass. Precisely the case, where we have equations between the co-ordinates of material points (forced motion), seems to me more an abstraction than a real case corresponding to nature. In all other cases, however, his definition begins to waver, as stands out especially when he introduces mass and density of volume elements for elastic bodies.

Hertz's mechanics perhaps becomes a perfectly clear, unambiguously determined picture through the fact that it knows no forces other than those attending forced motion. As Hertz puts it, his theory corresponds to the laws of thought. There is only one thing that I find lacking here, namely the proof that nature can really be represented by this picture.

Of course I am not denying that it is possible clearly to represent these concepts in ways other than mine; but it seems that it has never been achieved yet. Nor do I wish to be understood as saying that I find it unlikely that action at a distance between material points appearing as a function of distance must be the final picture of natural processes. Many attempts have been made to explain them by impacts of the molecules of a medium (the luminous aether), but this requires fairly intricate sub-sidiary assumptions, for example ascribing a lattice structure to ponderable molecules to explain cohesion. Moreover one needs the laws of impact and thus again the concept of mass. On a different tack, the attempt has been made to view molecules as vortex rings in order to explain thereby their apparent action at a distance. It is certainly possible that such attempts at explanation may one day displace forces at a distance. How-ever, the assumptions hitherto made to this end seem to me neither simpler nor clearer than the picture from which I started here. Rather, it seems to me that they uselessly increase the number of arbitrary hypotheses without corresponding gain in clarity and precision, which in my view must be avoided just as much as pictures that are unclear and vague because insufficiently specialised. In any case instead of asking how things are really constituted I should like to ask more modestly by means of what pictures our experience is currently most simply and unambig-uously represented.

However this may be, whether future efforts at perfecting mechanics are to be expected from further development of the special pictures current today or from their replacement by more general ideas of energetic or phenomenological character, at all events I believe that a clear and precise account of present day atomist mechanics can only be useful; for in the first case it will furnish the basis for further development and in the second it can serve as a model of that clarity and consistency indispensable for any new theory. These features seem to me by no means to lie in agreement with special laws of thought, but rather to rest on the fact that it uses only such rules and constructions as will always admit unambiguously defined application according to experience and, even where the outcome to be obtained is not known in advance, will furnish a clear and definite result in agreement with observation.

PREFACE TO PART II

PART II

> I first from Goethe mottoes chose,
> One of my own did next compose;
> Then Heine's poems I did con,
> But mottoes therein found I none.

PREFACE

In this second part of my lectures on mechanics I have set myself the task of treating the principle of least action, Hamilton's principles, and the connected work of Helmholtz, Hölder, Voss and others. However, the main emphasis is always on the physical sense and the connection with theorems of theoretical physics, not on purely mathematical deduction. As for additional comments required to remove all doubt concerning the mathematical rigour of the proofs, I refer to the writings of mathematicians; to me the main thing was always the physically intuitive aspect.

Special account had to be taken of the relations of the action principles to gas theory, heat, electricity and to the theorems of Maxwell, Helmholtz and Hertz on cyclical systems. However, I have nowhere gone into special areas of physics, but developed the subject just far enough for physics to continue from there.

In particular, Hamilton's principles of stationary and varying action presuppose no other knowledge except that of the total energy as function of all the variables on which it depends. That function being given, the principles enable us to derive all equations for all temporal changes that need be considered. This then amounts to the only unobjectionably founded theory of energetics that is unambiguously applicable without further ado in all cases.

The concept of mathematical variation seemed to me to become more intuitive by my connecting with it the concept of physical variation, the sequential ranging of infinitely many mathematical variations into a finite change of state, of the kind that is often used in thermodynamics for the treatment of the mechanical analogies of the second law. On the other hand the concept has also been contrasted with the theory of small finite disturbances in the method of variation of constants.

From Hamilton's principle of stationary action I derive Lagrange's equation for generalised co-ordinates, used for developing the general theory of the motion of rigid bodies, for introducing elliptical co-ordinates and deriving the theory of relative motion.

I owe important hints to Voss's article on this subject in the *Encyclopaedia of the Mathematical Sciences* and to talks on mechanics at the Congress of Scientists in Kassel and private conversations arising out of it both there and in Göttingen.

The usual formula for the period of oscillation of a conically swinging pendulum turns out to be valid if it always faces in the same direction in space. If it always faces the vertical axis of rotation, then besides the pendulum motion it has an additional rotation about its own longitudinal axis. If a pendulum contains a rotating body, the slow rotation of its plane of oscillation is opposite to the precessional motion, other things being equal. For the former, if regarded as made up of two superimposed equal and opposite conical pendulum swings of different period, will follow the faster, whereas precession is the limit approached by the slower when the elongation tends to 90°, while the period of the faster would grow to infinity.

In conclusion I hope that the strong physical leaning will not deter mathematicians from reading this book, nor the somewhat extensive formulae the physicist.

Vienna, June, 1904

§35. *The Principle of Action as the Fundamental Principle of all Natural Science*

Historically the idea of central forces between material points led either more openly or with concealment to the gradual development of mechanics in its present form. From this alone we must of course not infer that this idea is bound for ever to remain its basis. It happens often enough that a proposition that was first obtained under certain restrictive conditions later turns out to be valid in more general cases as well. Thus the principles of mechanics, such as that of virtual displacements or of stationary action might hold even under conditions that cannot be realised by central forces.

Indeed, the view has often been expressed that the idea of central forces should be completely dropped and be replaced by any one of the general principles as a basis for mechanics. If for this we choose the principle of energy, then since it is much more special than the equations of mechanics, we must add a whole series of other propositions and it is all over with the derivation of the whole material from a unitary principle. This would not be necessary if we chose the principle of stationary action, since from it the equations of mechanics do indeed follow in their entirety.

We might here even consider the case where the state of systems is determined by co-ordinates other than of position in three-dimensional space. For example Gibbs, Helmholtz and others have established relations in which temperature, the electric state and similar variables occur and which contain the mechanical principles, especially that of stationary action, as special cases. However, these relations greatly lack generality in other directions. They sometimes hold exclusively for states that differ only infinitesimally from equilibrium. Moreover they contain obscurities alien to mechanics, such as the concept of entropy, irreversibility and numerous empirically given properties of temperature, electricity and so on, the ideas of which are by no means so simple as those of geometrical relations of points.

Possibly, the appearance of equations analogous to the mechanical ones in electricity, thermodynamics and so on, as well as the special properties belonging to the magnitudes occurring in those theories, may be explained by assuming that these phenomena are caused by hidden mechanical motions and one might illuminate obscurities of behaviour of magnitudes occurring elsewhere in physics by mechanical pictures, for example obscurities in the concepts of entropy and irreversibility by applying calculus of probability to the behaviour of very numerous material points.

When I say that mechanical pictures might be able to illuminate such obscurities, I do not mean by this that the position and motion of material points in space is something whose simplest elements are completely explicable. On the contrary, to explain the ultimate elements of our cognition is altogether impossible; for to explain is to reduce to something better known and simpler, and therefore that to which everything is reduced must remain forever inexplicable. Thus even if everything were explained from the simplest fundamental concepts of mechanics, these would forever remain just as inexplicable as those of electricity are for us today.

Nor will I contend whether the concept of position in space, or of temperature, or of electric charge is clearer in itself: such quarrels would be empty. Only it would certainly be clearer if we could explain not only all phenomena of motion in solid, liquid and gaseous bodies but also heat, light, electricity, magnetism, and gravitation by means of the idea of motions of material points in space; that is, by means of a single unitary principle, instead of requiring for each of these agencies a whole inventory of quite alien concepts like temperature, electric charge, potential and so on, whether we denote these alien concepts as something quite independent or as disparate energy factors to be separately postulated for each energy form.

If one wants to bother at all about future centuries or even millennia I readily admit that it would be presumptuous to hope that our present day mechanical picture of the world will be preserved for all eternity even only in its most essential features.

Therefore I am far from despising attempts to find more general equations of which the mechanical equations are only special cases. Indeed I should be satisfied with the result of this book if, by showing how clear a picture of the world can and must be, I had contributed to the successful construction of another still more comprehensive and clearer world picture, whether on the basis of the energy principle or of the principle of stationary action or of the straightest path. I merely wish to work against the thoughtless attitude that declares the old world picture of mechanics an outworn point of view, before another such picture is available from its first foundations up to the application to the most important phenomena which the old picture has for so long now represented so exhaustively, especially where the innovators have not the least understanding of how difficult it is to construct such a picture. Above all, if one wants to avoid the picture of material points, one should not later introduce them into mechanics after all, but one should start from individuals or elements of different constitution,[8] with properties that can be described as clearly as those of material points.

I wrote the foregoing about seven years ago, the final sentence thus representing my requirements at that time (a measure of how old the basic stock of this book is). All this is now deliberately printed without change. Of what I then expected after centuries or even millennia, the half has happened in seven years.

However, the ray of hope for a non-mechanical explanation of nature came not from energetics or phenomenology, but from an atomic theory that in its fantastic hypotheses surpasses the old atomic theory as much as their elementary structures surpass in smallness those of the old atoms. I need not mention that I mean the modern theory of electrons. This certainly does not aim at explaining the concepts of mass, force, the law of inertia and so on from simpler and more easily understandable concepts and its simplest fundamental concepts and laws will doubtless remain just as inexplicable as those of mechanics for the mechanical picture of the world. However, the advantage of being able to derive all mechanics from other ideas that are in any case necessary for explaining electro-magnetism would be just as great as if conversely electro-magnetic phenomena could be explained mechanically. May the former succeed and my requirement of seven years ago be fulfilled!

§77. *Absolute and Relative Motion*

We can determine only the distances between different parts of bodies, that is only their relative position. There is no experience in which an absolute space would make itself felt. Nevertheless at the beginning of the book we introduced a definite co-ordinate system which almost plays the role of absolute space. We did this merely because by introducing this system we can state the laws of relative motion of bodies much more simply than by taking other systems, chosen quite arbitrarily.

By this we wish in no way to pronounce it likely, let alone necessary, that new empirical findings might supervene and enable us to determine this special co-ordinate system further, or to select a definite system from all those that we earlier (I §11) called suitable and thus to determine an absolute space; which, as the phrase goes, would prove the existence of absolute space.

For we saw above (I, §11) that this reduction of the laws of motion to their simplest form can be carried out not just on the basis of one single and definite system S, but that one may equally use very different systems. All these we there called suitable reference frames. In this the direction of the axes in space at a definite moment of time and the position of the origin for two moments of time can be quite arbitrarily orientated relatively to the systems already found to be suitable. If, however, one has chosen the direction of the axes at a moment, they will be thereby fixed at all other times. All directions that a definite axis then always has are called parallel.

If moreover the position of the origin at two moments has been chosen, its position at all other times is again determined. The motion of the origin under these conditions is called rectilinear and uniform.

The question as to how the laws of a body's change of position are modified, that is how the equations of motion alter if at different times we base ourselves on co-ordinate systems not obeying these conditions, is evidently of great theoretical interest.

It has, however, practical value too; for we only ever observe the relative motion of one material system with regard to a second that is quasi-immutable or at least is regarded as immutable. Thus we observe the motion of the planetary system relative to the sphere of fixed stars, that of terrestrial bodies relative to the Earth or some other object rigidly fixed to it. In certain experiments we observe the motion of fluids or other objects relative to a vessel or chamber deliberately set into rotational motion. People in a moving carriage or ship can observe the motion of their bodies and other objects relative to the carriage or ship and so on.

In all these cases it is for us a matter only of relative motion of the first system with regard to the second or a co-ordinate system rigidly fixed to the latter. In all cases except the first, this co-ordinate system certainly lacks the properties of a suitable reference frame. The nature of the fixed stars is much too unknown and the firmament itself much too indefinite a concept for us to be able to decide with certainty whether a co-ordinate system rigidly fixed to it is suitable; however, proper motions of fixed stars themselves have already been observed and at all events it is important in that case too what influence it would have on the equations of motion of the planetary system if the co-ordinate system on which they are based was not a suitable reference frame.

If we are to calculate the motion of a system of bodies relative to a second and the latter's motion relative to a suitable reference frame is known, we could in each special case proceed as follows; first calculate the motion of the first system of bodies relative to the reference frame and only then calculate the motion of the first system relative to the second from that of each relative to the common suitable reference frame.

However it is highly advantageous not to work this out separately in each special case but to give once and for all the rules from which we can directly find the relative motion of one system of bodies with regard to the second or a co-ordinate system rigidly fixed to it as soon as we are given the motion of the second relative to a suitable reference frame that we call the co-ordinate system at rest. We assume that all parts of the second system are rigidly connected with each other, and we now imagine a second co-ordinate system rigidly connected with that system of bodies (the moving co-ordinate system). The problem then is to find the general equations of motion for the first system of bodies relative to the moving co-ordinate system.

§88. *The Law of Inertia*

In conclusion let us return to a fundamental difficulty in the simplest fundamental laws of mechanics, namely the formulation of the law of inertia, if one does not wish to introduce an absolute transcendental space. We have evaded this difficulty in the simplest way by never speaking of anything real or existing, but replacing matter by mere mental pictures, namely material points, without worrying whether this might not also be done equally successfully in other ways (for example on the basis of some other co-ordinate system).

No one can stop us forming mental pictures as we wish, nor therefore including in them a co-ordinate system (the suitable reference frame) over and above the material points. After the event we call these mental pictures true only because they are useful in predicting future phenomena (our future sensations) as completely and effortlessly as possible, that is in adapting our volitional impulses to them.

I can find no inner reason for the assertion that mental pictures containing the reference frame cannot agree exactly with experience. On the contrary, I find that it is precisely this assertion that aims at stating something a priori about experience which the latter did not tell us before. I therefore cannot here agree with Mach; if I understand him correctly, he infers from the fact of our thoughts' having to represent only relations between objects that the law of inertia could be determined only by the world of fixed stars. If the picture we have developed here were absolutely correct, then the relations between objects would be such that a mental reference frame of this kind is required to represent them in the simplest way. It would not be necessary for the firmament to be absolutely free of rotation relative to this reference frame; that in fact it is nearly so, would be explicable from their great distance, at which rotation would require enormous centripetal forces.

For somebody who surveyed the whole world of fixed stars, imagined as finite, it would even be theoretically possible to observe its rotation by means of Foucault's pendulum or a gyroscope as we could do for the Earth if we lacked light and therefore knowledge of other celestial bodies. However, this observation would not be unconditional, though if it were denied one would have to make an assumption much more unnatural still and running counter to the presuppositions of this book, namely of a force exerted from a fixed straight line on all masses, proportional to mass and distance, and of the corresponding Coriolis force which further depends on velocity and position in space relative to the firmament. Nor would a single Foucault pendulum or gyroscope suffice, for in that case the possibility of such small disturbing forces could hardly be ruled out with certainty. However, if the most varied Foucault pendulums, Streintz gyroscopes, material points propelled according to Lange and so on, all indicate by their motion relative to the world of stars (taken as finite and observable as a whole) that the latter is rotating, we could surely take it as most likely that by assuming such a rotation (that is by adopting a reference frame into our mental picture), phenomena are most simply explicable, or perhaps we should rather say describable or mentally repeatable; just as we are already convinced that only by assuming a rotation of the Earth can cosmic phenomena be rationally described.

That there is a real body α, always at rest relative to our suitable reference frame and able to replace the latter, would be an absurd idea. As mere mental object, however, I prefer to call it 'reference frame' than 'body α'. The name 'body' alone seems to me most inappropriate.

Let us therefore assume that the pictures described in this book exactly represent nature and that the world is finite. In that case its invariable axis through the centre of gravity with regard to every suitable reference frame would have to retain its direction without change and the total areal moment of the world with regard to that axis would have to be constant. However, this merely means the following: what is given is only the relative positions of all parts of the world at all times. We can relate them to an arbitrary co-ordinate system chosen differently at any time, and likewise arbitrarily call any two stretches of time equally long or not; but only a definite way of denoting stretches of time as equal and only a definite temporal sequence of positions of these co-ordinate systems fulfils the condition that changes of material points of the world with regard to these requirements obey the simple equations of mechanics expounded in this book. Only if this is the case do we call the temporal sequence of the co-ordinate systems in question a suitable reference frame.

For each such suitable reference frame what has been proved shows that that axis through the world's centre of gravity for which the sum of areal moments of the world is a maximum must have an invariable position and this sum itself be constant. This does of course further presuppose that we are going to think of the world as being finite. It would, however, be quite possible for the different equations of mechanics for such a sequence of positions of the co-ordinate system to take on their simple form for which the constant areal moment of the world about its invariable axis through the centre of gravity is different from zero, so that the world would as it were rotate, though of course at a rate that would be small in the same measure as the world is big.

Though it is totally unlikely that our knowledge will ever really extend that far, we can nevertheless imagine ourselves in the position where we know the whole world of fixed stars (provided we take it as finite) as well as we know our Earth today and where we could establish a rotation of the former as securely as we can that of the latter. If light did not exist so that we were ignorant of any celestial bodies except the Earth, we should doubtless have reached the concept of absolute rotation much later. However, we could certainly have reached it by experiments with gyroscopes, pendulums and so on and would very likely have done so. It is thus by no means just the relation to the firmament that conditions this concept.

That the successive positions of the principal inertial axes of the world relative to its centre of gravity form a suitable reference frame of this kind would of course be possible and by no means in conflict with our past experience, according to which any axes immutably connected with the nearly immutable firmament are suitable reference frames. On this assumption it would be superfluous to adopt a special co-ordinate system, since its position could be calculated for any moment of time from the successive relative positions of all parts of the world. Since, however, it has not been and cannot be proved by any considerations that such an assumption is correct, it is of course well to point out that it may be so but it would evidently be an illicit restriction of generality to place the assumption at the head of mechanics.

Quite independently of this there is the question whether the mechanical equations here developed and therefore also the law of inertia might perhaps be only approximately correct and whether, by formulating them more correctly, the improbability or rather inhomogeneity of having to adopt into the picture a co-ordinate system as well as material points would disappear of itself.

Here Mach pointed to the possibility of a more correct picture, obtained by assuming that only the acceleration of the change of distance between any two material particles is determined mainly by the neighbouring masses, its velocity being determined by a formula in which very distant masses are decisive. This naturally avoids the adopting of any co-ordinate system into the picture, since now it is only a question of distances. Of course, Mach does not avoid introducing other difficulties, for example that the world is finite, a kind of action at a distance for the greatest distances and so on. These difficulties did not seem to me to be so particularly great as perhaps to some other physicists, but they certainly have the awkward feature that they seem to exclude all empirical test forever. Nor do I think that it is obvious that we can give an exact derivation of the principle of superposition by continual application of the new form of the law of inertia to the centre of gravity of the masses in mutual interaction. Moreover

$$\frac{d^2 \sum mr}{dt^2} = 0,$$

is merely an equation and thus not equivalent to the assertion that a point moves uniformly in a straight line.

At all events I think that such an extension of our vision, by pointing out that what we regard as most certain and obvious may perhaps be only approximately correct, is most valuable. It is in line with the suggestion that the distances of fixed stars may perhaps be constructed only in a non-Euclidean space of very small curvature, which is of course connected with the law of gravity in that a moving body not acted on by forces would then after aeons have to return to its previous position if the curvature in question is positive.

In all these considerations we started from the presupposition that the world is finite. If one conceives the world as infinite, concepts such as the world's centre of gravity, invariable axis, principal inertial axes and so on become quite empty. One would then have to assume that the law of inertia is determined by a formula according to which masses that are nearby have vanishing influence on the formulation of the law of inertia, that those at distances like Sirius have the greatest such influence and those at much greater distance still again next to none.

All difficulties in formulating the law of inertia are avoided by the electromagnetic theory of matter, which assumes that Maxwell's equations for the behaviour of the luminous aether and the motions of electrons in it are the primary notions from which follow the law of inertia for the motion of electrons relative to the luminous aether and the other mechanical laws for such motion. However, the law of inertia does not hold for the particles of the aether itself; Maxwell's equations would have to be formulated in such a way that they determine only the mutual actions of adjacent volume elements so that we need no absolute space to formulate them. A working out of this as yet quite undeveloped theory is no concern of ours here.

NOTES

[1] Heinrich Hertz, *Principles of Mechanics*, 1894, English edn. 1899.

[2] Cf. Boltzmann, *Wien. Ber.* **105** (1896) 907; and the essay on p. 41 above.

[3] Here a generalisation of the picture is possible without serious loss of clarity, and has indeed been tried, by assuming that the function F contains the first or even second time derivative of r_{12}. If the latter figures in it linearly one might equally well say that the factors m, to be mentioned later, are not constant. However, this generalisation has gained so little practical importance that we shall refrain from discussing it further here.

[4] *Carls, Repertorium* **4** (1868) 355–9

[5] Cf. Boltzmann, *loc. cit.*

[6] Kirchhoff's well-known demand that physics is merely to describe facts is satisfied by our picture insofar as it is merely a set of rules for constructing arithmetical and geometrical ideas by means of which the facts can always be correctly predicted. The concepts of cause and effect are entirely avoided in this. For even if one occasionally denotes the presence of one material point as the cause of the other's acceleration, this is merely to express the idea of the fact that both receive certain accelerations at a certain mutual distance from each other. I therefore hope that there can be no epistemological objections to the mode of presentation adopted here.

[7] That any clear picture of the mutual action of volume elements must have much more in common with current atomism than is ordinarily assumed I believe I have established elsewhere. In the same place I have shown that it confuses the idea and makes calculation of the limit impossible if one assumes that the simple elements are themselves extended and divisible into differentials. Cf. Boltzmann, *loc. cit.*

[8] Cf. the essay on p. 41 above.

BIBLIOGRAPHY

I. SCIENTIFIC AND PHILOSOPHICAL
WRITINGS OF L. BOLTZMANN

The following list is based on the bibliography given in F. Hasenöhrl's collection of Boltzmann's *Wissenschaftliche Abhandlungen*. Like that bibliography it omits ephemera – newspaper articles, notices of books, etc. – which are now mostly untraceable. The writings are arranged by year of publication or (in the case of addresses) delivery. Within each year books are listed first, and a roughly thematic rather than a chronological order is followed for articles, lectures, etc. Contemporary or subsequent English translations (including those in the present volume) are indicated in **bold type**.

Abbreviations are generally readily intelligible. *Sitzungsberichte* of the various academies (*mathematisch-naturwissenschaftliche* or *mathematisch-physikalische Klasse*) are indicated by '*Wien. Ber.*', '*Münch. Ber.*', and the like. The reprinting of a piece in *Wissenschaftliche Abhandlungen* or *Populäre Schriften* is indicated by '= WA I.1' (giving the volume and number of the piece) or '= PS 4, pp. 51–75' (giving the number of the piece and the page references).

1865　Über die Bewegung der Elektrizität in krummen Flächen (*Wien. Ber.* **52** (1865) 214–221 = WA I.1)

1866　Über die mechanische Bedeutung des zweiten Hauptsatzes der Wärmetheorie (*Wien. Ber.* **53** (1866) 195–220 = WA I.2)

1867　Über die Anzahl der Atome in den Gasmolekülen und die innere Arbeit in Gasen (*Wien. Ber.* **56** (1867) 682–690 = WA I.3)

1868　Über die Integrale linearer Differentialgleichungen mit periodischen Koeffizienten (*Wien. Ber.* **58** (1868) 54–59 = WA I.4)
Studien über das Gleichgewicht der lebendigen Kraft zwischen bewegten materiellen Punkten (*Wien. Ber.* **58** (1868) 517–560 = WA I.5)
Lösung eines mechanischen Problems (*Wien. Ber.* **58** (1868) 1035–1044 = WA I.6)

1869　Über die Festigkeit zweier mit Druck übereinandergesteckter zylindrischer Röhren (*Wien. Ber.* **59** (1869) 679–688 = WA I.7)
Über die elektrodynamische Wechselwirkung der Teile eines elektrischen Stromes von veränderlicher Gestalt (*Wien. Ber.* **60** (1869) 69–87; *Schlömilchs Z. S.* **15** (1870) 16ff. = WA I.8)
Bemerkung zur Abhandlung des Hrn. R. Most: Ein neuer Beweis des zweiten Wärmegesetzes (*Pogg. Ann.* **137** (1869) 495 = WA I.9)

1870 Erwiderung an Hrn. Most
(*Pogg. Ann.* **140** (1870) 635–644 = *WA* I.10)
Über die von bewegten Gasmassen geleistete Arbeit
(*Pogg. Ann.* **140** (1870) 254–263 = *WA* I.11)
Noch Einiges über Kohlrauschs Versuch zur Bestimmung des Verhältnisses der Wärmekapazitäten von Gasen
(*Pogg. Ann.* **141** (1870) 473–476 = *WA* I.12)
Über die Ableitung der Grundgleichungen der Kapillarität aus dem Prinzipe der virtuellen Geschwindigkeiten
(*Pogg. Ann.* **141** (1870 582–590 = *WA* I.13)
Über eine neue optische Methode, die Schwingungen tönender Luftsäulen zu analysieren, gemeinschaftlich mit A. Toepler
(*Pogg. Ann.* **141** (1870) 321–352 = *WA* I.14)
1871 Boiling-Points of Organic Bodies
(*Philosophical Magazine and Journal* **42** (1871) 393 = *WA* I.15, **English original**)
Über die Druckkräfte, welche auf Ringe wirksam sind, die in bewegte Flüssigkeit tauchen
(*Crelle Journ.* **73** (1871) 111–134 = *WA* I.16)
Zur Priorität der Auffindung der Beziehung zwischen dem zweiten Hauptsatze der mechanischen Wärmetheorie und dem Prinzip der kleinsten Wirkung
(*Pogg. Ann.* **143** (1871) 211–230 = *WA* I.17)
Über das Wärmegleichgewicht zwischen mehratomigen Gasmolekülen
(*Wien. Ber.* **63** (1871) 397–418 = *WA* I.18)
Einige allgemeine Sätze über Wärmegleichgewicht
(*Wien. Ber.* **63** (1871) 679–711 = *WA* I.19)
Analytischer Beweis des zweiten Hauptsatzes der mechanischen Wärmetheorie aus den Sätzen über das Gleichgewicht der lebendigen Kraft
(*Wien. Ber.* **63** (1871) 712–732 = *WA* I.20)
1872 Über das Wirkungsgesetz der Molekularkräfte
(*Wien. Ber.* **66** (1872) 213–219 = *WA* I.21)
Weitere Studien über das Wärmegleichgewicht unter Gasmolekülen
(*Wien. Ber.* **66** (1872) 275–370 = *WA* I.22)
Resultate einer Experimentaluntersuchung über das Verhalten nicht leitender Körper unter dem Einflusse elektrischer Kräfte
(*Wien. Ber.* **66** (1872) 256–263 = *WA* I.23)
1873 Experimentelle Bestimmung der Dielektrizitätskonstante von Isolatoren
(*Wien. Ber.* **67** (1873) 17–80; *Pogg. Ann.* **151** (1874) 482 and 531; *Carls Repertorium* **10** (1874) 109 = *WA* I.24)
Experimentaluntersuchung über die elektrostatische Fernwirkung dielektrischer Körper
(*Wien. Ber.* **68** (1873) 81–155 = *WA* I.25)
1874 Experimentelle Bestimmung der Dielektrizitätskonstante einiger Gase
(*Wien. Ber.* **69** (1874) 795–813; *Pogg. Ann.* **155** (1875) 403 = *WA* I.26)
Über einige an meinen Versuchen über die elektrostatische Fernwirkung dielektrischer Körper anzubringende Korrektionen
(*Wien. Ber.* **70** (1874) 307–341 = *WA* I.27)
Über die Verschiedenheit der Dielektrizitätskonstante des kristallisierten Schwefels nach verschiedenen Richtungen

(*Wien. Ber.* **70** (1874) 342–366 = *WA* I.28)
Experimentaluntersuchung über das Verhalten nicht leitender Körper unter dem Einflusse elektrischer Kräfte
(*Pogg. Ann.* **153** (1874) 525–534 = *WA* I.29: extract from detailed papers in *Wien. Ber.* **66** (1872), **68** (1873), and **70** (1874))
Zur Theorie der elastischen Nachwirkung
(*Wien. Ber.* **70** (1874) 275–306; *Pogg. Ann. Erg.-Bd.* **7** (1876) 624 = *WA* I.30)
Über den Zusammenhang zwischen der Drehung der Polarisationsebene und der Wellenlänge der verschiedenen Farben
(*Pogg. Ann. Jubelband* (1874) 128–134 = *WA* I.31)
1875 Über das Wärmegleichgewicht von Gasen, auf welche äußere Kräfte wirken
(*Wien. Ber.* **72** (1875) 427–457 = *WA* II.32)
Bemerkungen über die Wärmeleitung der Gase
(*Wien. Ber.* **72** (1875) 458–470; *Pogg. Ann.* **157** (1876) 457–469 = *WA* II.33)
Zur Integration der partiellen Differentialgleichungen I. Ordnung
(*Wien. Ber.* **72** (1875) 471–483 = *WA* II.34)
1876 Zur Abhandlung des Hrn. Oscar Emil Meyer über innere Reibung
(*Crelle Journal* **81** (1876) 96 = *WA* II. 35)
Über die Aufstellung und Integration der Gleichungen, welche die Molekularbewegung in Gasen bestimmen
(*Wien. Ber.* **74** (1876) 503–552 = *WA* II. 36)
Über die Natur der Gasmoleküle
(*Wien. Ber.* **74** (1876) 553–560 = *WA* II.37)
Zur Geschichte des Problems der Fortpflanzung ebener Luftwellen von endlicher Schwingungsweite
(*Schlömilchs Zeitschrift* **21** (1876) 452 = *WA* II.38
1877 Bemerkungen über einige Probleme der mechanischen Wärmetheorie
(*Wien. Ber.* **75** (1877) 62–100 = *WA* II.39)
Notiz über die Fouriersche Reihe
(*Wien. Anz.* **14** (1877) 10 = *WA* II.40)
Über eine neue Bestimmung einer auf die Messung der Moleküle Bezug habenden Größe aus der Theorie der Kapillarität
(*Wien. Ber.* **75** (1877) 801–813 = *WA* II.41)
Über die Beziehung zwischen dem zweiten Hauptsatze der mechanischen Wärmetheorie und der Wahrscheinlichkeitsrechnung respektive den Sätzen über das Wärmegleichgewicht
(*Wien. Ber.* **76** (1877) 373–435 = *WA* II. 42)
Über einige Probleme der Theorie der elastischen Nachwirkung und über eine neue Methode Schwingungen mittels Spiegelablesung zu beobachten, ohne den schwingenden Körper mit einem Spiegel von erheblicher Masse zu belasten
(*Wien. Ber.* **76** (1877) 815–842 = *WA* II. 43)
1878 Weitere Bemerkungen über einige Probleme der mechanischen Wärmetheorie
(*Wien. Ber.* **78** (1878) 7–46 = *WA* II.44)
Über die Beziehung der Diffusionsphänomene zum zweiten Hauptsatze der mechanischen Wärmetheorie
(*Wien. Ber.* **78** (1878) 733–763 = *WA* II.45
Zur Theorie der elastischen Nachwirkung
(*Wied. Ann.* **5** (1878) 430–432 = *WA* II.46
Remarques au sujet d'une Communication de M. Maurice Lévy, sur une loi

universelle relative à la dilatation des corps
   (*C. R.* **87** (1878) 593 = *WA* II.47)
Nouvelles remarques au sujet des Communications de M. Maurice Lévy, sur
une loi universelle relative à la dilatation des corps
   (*C. R.* **87**(1878) 773 = *WA* II.48)
Notiz über eine Arbeit des Hrn. Oberbeck über induzierten Magnetismus
   (*Wien. Anz.* **15** (7 November 1878 = *WA* II.49) 203–205)
1879   Über das Mitschwingen eines Telephons mit einem anderen
   (*Wien. Anz.* **16** (March 1879) 71–73 = *WA.* II.50)
Über die auf Diamagnete wirksamen Kräfte
   (*Wien. Ber.* **80** (1879) 687–714 = *WA* II.51)
Erwiderung auf die Bemerkung des Hrn. Oskar Emil Meyer
   (*Wien. Ann.* **8** (1879) 653–655 = *WA* II.52)
1880   Erwiderung auf die Notiz des Hrn. O. E. Meyer: „Über eine veränderte Form"
usw.
   (*Wien. Ann.* **11** (1880) 529–534 = *WA* II.53)
Über die Magnetisierung eines Ringes. Über die absolute Geschwindigkeit der
Elektrizität im elektrischen Strome
   (*Wien. Anz.* **17** (15 January 1880) 12–13 = *WA* II.54; *Phil. Mag.* (5) **9**,
   307–309, in English translation)
Zur Theorie der sogenannten elektrischen Ausdehnung oder Elektrostriktion I
   (*Wien. Ber.* **82** (1880) 826–839 = *WA* II.55)
Zur Theorie der sogenannten elektrischen Ausdehnung oder Elektrostriktion II
   (*Wien. Ber.* **82** (1880) 1157–1168 = *WA* II.56)
Zur Theorie der Gasreibung I
   (*Wien. Ber.* **81** (1880) 117–158 = *WA* II.57)
1881   Zur Theorie der Gasreibung II
   (*Wien. Ber.* **84** (1881) 40–135 = *WA* II.58)
Zur Theorie der Gasreibung III
   (*Wien. Ber.* **84** (1881) 1230–1263 = *WA* II.59)
Entwicklung einiger zur Bestimmung der Diamagnetisierungszahl nützlichen
Formeln
   (*Wien. Ber.* **83** (1881) 576–587 = *WA* II.60)
Einige Experimente über den Stoß von Zylindern
   (*Wien. Ber.* **84** (1881) 1225–1229; *Wied. Ann.* **17** (1882) 343–347 = *WA* II.61)
Über einige das Wärmegleichgewicht betreffende Sätze
   (*Wien. Ber.* **84** (1881) 136–145 = *WA* II.62)
Referat über die Abhandlung von J. C. Maxwell „Über Boltzmanns Theorem
betreffend die mittlere Verteilung der lebendigen Kraft in einem System ma-
terieller Punkte")
   (*Wied. Ann. Beiblätter* **5** (1881) 403–417 = *WA* II. 63; *Phil. Mag.* (5) **14**
   (1882) 299–413, in English translation)
Zu K. Streckers Abhandlung: Über die spezifische Wärme des Chlors usw.
   (*Wied. Ann.* **13** (1881) 544 = *WA* II.64)
1882   Vorläufige Mitteilung über Versuche, Schallschwingungen direkt zu photo-
graphieren
   (*Wien. Anz.* **19** (30 November 1882) 242–243 = *WA* III.65; *Phil. Mag.*
   (5) **15** (1883) 151, in English translation)
Zur Theorie der Gasdiffusion I

(*Wien. Ber.* **86** (1882) 63–99 = *WA* III.66)

ed.: G. R. Kirchhoff, *Gesammelte Abhandlungen, Nachtrag* (J. A. Barth, Leipzig, 1882)

1883   Zur Theorie der Gasdiffusion II

(*Wien. Ber.* **88** (1883) 835-860 = *WA* III.67)

Zu K. Streckers Abhandlungen: Die spezifische Wärme der gasförmigen zweiatomigen Verbindungen von Chlor, Brom, Jod usw.

(*Wied. Ann.* **18** (1883) 309–310 = *WA* III.68)

Über das Arbeitsquantum, welches bei chemischen Verbindungen gewonnen werden kann

(*Wien. Ber.* **88** (1883) 861–896; *Wied. Ann.* **22** (1884) 39–72 = *WA* III.69

1884   Über die Möglichkeit der Begründung einer kinetischen Gastheorie auf anziehende Kräfte allein

(*Wien. Ber.* **89** (1884) 714–722; *Wied. Ann.* **24** (1885) 37–44; *Exner Rep.* **21** (1885) 1–7 = *WA* III.70)

Über eine von Hrn. Bartoli entdeckte Beziehung der Wärmestrahlung zum zweiten Hauptsatze

(*Wied. Ann.* **22** (1884) 31–39 = *WA* III.71)

Ableitung des Stefanschen Gesetzes, betreffend die Abhängigkeit der Wärmestrahlung von der Temperatur aus der elektromagnetischen Lichttheorie

(*Wied. Ann.* **22** (1884) 291–294 = *WA* III.72)

Über die Eigenschaften monozyklischer und anderer damit verwandter Systeme

(*Crelles Journal* **98** (1884 and 1885) 68–94 = *WA* III.73)

1885   Über einige Fälle, wo die lebendige Kraft nicht integrierender Nenner des Differentials der zugeführten Energie ist

(*Wien. Ber.* **92** (1885) 853–875; *Exners Rep.* **22** (1886) 135–154 = *WA*. III.74

1886   Neuer Beweis eines von Helmholtz aufgestellten Theorems betreffend die Eigenschaften monozyklischer Systeme

(*Göttinger Nachrichten* (1886) 209–213 = *WA* III.75)

Notiz über das Hallsche Phänomen

(*Wien. Anz.* **23** (8 April 1886) 77–80 = *WA* III.76; *Phil. Mag.* (5) **22** (1886) 226–228 **in English translation**)

Zur Theorie des von Hall entdeckten elektromagnetischen Phänomens

(*Wien. Ber.* **94** (1886) 644–669 = *WA* III. 77)

Bemerkung zu dem Aufsatze des Hrn. Lorberg über einen Gegenstand der Elektrodynamik

(*Wied. Ann.* **29** (1886) 598–603 = *WA* III.78)

Über die von Pebal in seiner Untersuchung des Euchlorins verwendeten unbestimmten Gleichungen

(*Ann. d. Chemie* **232** (1886) 121–124 = *WA* III.79)

Zur Berechnung der Beobachtungen mit Bunsens Eiskalorimeter

(*Ann. d. Chemie* **232** (1886) 125–128 = *WA* III. 80)

Über die zum theoretischen Beweise des Avogadroschen Gesetzes erforderlichen Voraussetzungen

(*Wien. Ber.* **94** (1886) 613–643 = *WA* III.81; *Phil. Mag.* (5) **23** (1887) 305–333 **in English translation**)

1887   Über die mechanischen Analogien des zweiten Hauptsatzes der Thermodynamik

(*Crelles Journal* **100** (1887) 201–212 = *WA* III. 82)

Neuer Beweis zweier Sätze über das Wärmegleichgewicht unter mehratomigen Gasmolekülen
(*Wien. Ber.* **95** (1887) 153–164 = *WA* III.83)
Versuch einer theoretischen Beschreibung der von Prof. Albert v. Ettingshausen beobachteten Wirkung des Magnetismus auf die galvanische Wärme
(*Wien. Anz.* **24** (17 March 1887) 71–74 = *WA* III.84)
Über einen von Prof. Pebal vermuteten thermochemischen Satz, betreffend nicht umkehrbare elektrolytische Prozesse
(*Wien. Ber.* **95** (1887 = *WA* III. 85) 935–941)
Über einige Fragen der kinetischen Gastheorie
(*Wien. Ber.* **96** (1887) 891–918 = *WA* III.86; *Phil. Mag.* (5) **25** (1888) 81–103, **in English translation**)
Zur Theorie der thermoelektrischen Erscheinungen
(*Wien. Ber.* **96** (1887) 1258–1297 = *WA* III.87)
Einige kleine Nachträge und Berichtigungen
(*Wied. Ann.* **31** (1887) 139–140 = *WA* III.88)
Über die Wirkung des Magnetismus auf elektrische Entladungen in verdünnten Gasen
(*Wied. Ann.* **31** (1887) 789–792 = *WA* III.89)
Gustav Robert Kirchhoff
Address given at Graz University on 15 November 1887 (J. A. Barth, Leipzig (1888) = *PS* **4**, pp. 51–75)

1888   Über das Gleichgewicht der legendigen Kraft zwischen progressiver und Rotationsbewegung bei Gasmolekülen
(*Berl. Ber.* (1888) 1395–1408 = *WA* III. 90)

1889   Über das Verhältnis der Größe der Moleküle zu dem von den Valenzen eingenommenen Raume
(Lecture delivered to the 62. Versammlung D. Naturforscher und Ärzte in Heidelberg (1889)). Report by W. Nernst = *WA* III.91

1890   Über die Hertzschen Versuche
(*Wied. Ann.* **40**, 399–400 = *WA* III.92; *Phil. Mag.* (5) **30** (1890) **126, in English translation**)
Die Hypothese van 't Hoffs über den osmotischen Druck vom Standpunkte der kinetischen Gastheorie
(*Zeitschrift f. phys. Chemie* **6** (1890) 474–480 = *WA* III.95)
Über die Bedeutung von Theorien
(Farewell address at Graz delivered on 16 July 1890 = *PS*, pp. 76–80, **translated in the present volume**)

1891   *Vorlesungen über die Maxwellsche Theorie der Elektrizität und des Lichtes, I. Theil*
(J. A. Barth, Leipzig, 1891)
Nachtrag zur Betrachtung der Hypothese van 't Hoffs vom Standpunkte der kinetischen Gastheorie
(*Zeitschr. f. phys. Chemie* **7** (1891) 88–90 = *WA* III,94
Über einige die Maxwellsche Elektrizitätstheorie betreffende Fragen (*Verh. d. 64. Vers. D. Naturf. u. Ärzte*, Halle a. S. 1891, 29–34; *Wied. Ann.* **48** (1893) 100–107 = *WA* III.95

1892   Über ein Medium, dessen mechanische Eigenschaften auf die von Maxwell für den Elektromagnetismus aufgestellten Gleichungen führen

(*Münch. Ber.* 22 (1892) 279–301; *Wied. Ann.* 48 (1893) 78–99 = *WA* III.96)
III. Teil der Studien über Gleichgewicht der lebendigen Kraft
  (*Münch. Ber.* 22 (1892) 329–358 = *WA* III. 97; *Phil. Mag.* (5) 35 (1893) 153–173, in English translation)
Über ein mechanisches Modell zur Versinnlichung der Anwendung der Lagrangeschen Bewegungsgleichungen in der Wärme- und Elektrizitätslehre
  (*Jahresbericht d. Deutsch. Math.-Vereinigung* 1 (1892) 53–55 = *WA* III.98)
Beschreibung einiger Demonstrationsapparate
  (Deutsche Mathematiker-Vereinigung, Katalog mathem. Modelle etc. (1892) = *WA* III.99)
Über das den Newtonschen Farbenringen analoge Phänomen beim Durchgang Hertzscher elektrischer Planwellen durch planparallele Metallplatten
  (*Münch. Ber.* 22 (1892) 53–70; *Wied. Ann.* 48 (1893) 63–77 = *WA* III.100)
Über die Methoden der theoretischen Physik
  (Deutsche Mathematiker-Vereinigung, Katalog mathem. Modelle etc. (1892) = *PS* 1, 1–10, translated in the present volume)
1893 Vorlesungen über die Maxwellsche Theorie der Elektrizität und des Lichtes, II. Theil
  (J. A. Barth, Leipzig, 1893)
Über die Beziehung der Äquipotentiallinien und der magnetischen Kraftlinien
  (*Münch. Ber.* 23 (1893) 119–127; *Wied. Ann.* 51 (1894) 550–558 = *WA* III. 101)
Über die Bestimmung der absoluten Temperatur
  (*Münch. Ber.* 23 (1893) 321–328; *Wied. Ann.* 53 (1894) 948–954 = *WA* III.102)
Der aus den Sätzen über Wärmegleichgewichtfolgende Beweis des Prinzips des letzten Multiplikators in seiner einfachsten Form.
  (*Math. Ann.* 42 (1893) 374–376 = *WA* III.103)
Über die Notiz des Hrn. Hans Cornelius bezüglich des Verhältnisses der Energien der fortschreitenden und inneren Bewegung der Gasmoleküle
  (*Zeitschr. f. phys. Chemie* 11 (1893) 751–752 = *WA* III.104)
Über die neueren Theorien der Elektrizität und des Magnetismus
  (*Verh. d. 65. Vers. Deutsch. Naturf. u. Ärzte* (Nürnberg, 1893) 34–35 = *WA* III.105)
1894 Zur Integration der Diffusionsgleichung bei variablen Diffusionskoeffizienten
  (*Münch. Ber.* 24, 211–217; *Wied. Ann.* 53 (1894) 959–964 = *WA* III.106)
Über die mechanische Analogie des Wärmegleichgewichtes zweier sich beruhrender Körper. Gemeinsam mit G. H. Bryan
  (*Wien. Ber.* 103 (1894) 1125–1134 = *WA* III.107; *Proc. Phys. Soc. London* 13 (1895) 485–493, in English translation)
On the Application of the Determinantal Relation to the Kinetic Theory of Polyatomic Gases
  (Appendix C, G. H. Bryan's article on Thermodynamics II, *Rep. British Ass. for the Advanc. of Science*, Oxford, 1894, 102–106, = *WA* III.108 English original)
On Maxwell's Method of Deriving the Equations of Hydrodynamics from the Kinetic Theory of Gases
  (*Report of the British Association.* Oxford, 1894, 579 = *WA* III.109, English original)

Über den Beweis des Maxwellschen Geschwindigkeitsverteilungsgesetzes unter Gasmolekülen
    (*Münch. Ber.* **24**, 207–210; *Wied. Ann.* **53** (1894) 955–958 = *WA* III.110)
Über Luftschiffahrt
    (Lecture delivered to the Gesellschaft Deutscher Naturforscher und Ärzte, Vienna 1894 = *PS* 6, pp. 81–91)
1895 Nochmals das Maxwellsche Verteilungsgesetz der Geschwindigkeiten
    (*Münch. Ber.* **25**, 25–26; *Wied. Ann.* **55** (1895) 223–224 = *WA* III.111
On Certain Questions of the Theory of Gases
    (*Nature* **51** (1895) 413–415 = *WA* III.112; English original reprinted in the present volume)
Reply to Culverwell
    (*Nature* **51** (1895) 581 = *WA* III. 113, English original)
On the Minimum Theorem in the Theory of Gases
    (*Nature* **52** (1895) 221 = *WA* III.114; English original)
Josef Stefan
    (Address given in Vienna on 8 December 1895 = *PS* 7, pp. 92–103)
Zur Erinnerung an Josef Loschmidt
    (address given in Vienna on 29 October 1895 = *PS* 15, pp. 228–252)
    ed.: J. C. Maxwell, *Über Faradays Kraftlinien*, in Ostwalds Klassiker der exakten Wissenschaften, Leipzig
1896 *Vorlesungen über Gastheorie I. Theil*
    (J. A. Barth, Leipzig, 1896: translated by S. G. Brush as *Lectures on Gas Theory*, Berkeley, 1964)
Über die Berechnung der Abweichungen der Gase vom Boyle-Charlesschen Gesetz und der Dissoziation derselben
    (*Wien. Ber.* **105** (1896) 695–706 = *WA* III.115)
Ein Vortrag über die Energetik
    (Berichte über die Sitzungen der Chemisch-physikalischen Gesellschaft in Wien, 11 February 1896. *Vierteljahresber. d. Wiener Ver. z. Förderung des phys. u. chem. Unterrichts* 2, 38 = *WA* III.116)
Sur la théorie des gaz. Lettre à M. Betrand I.
    (*C. R.* **122** (1896) 1173 = *WA* III.117)
Sur la théorie des gaz. Lettre à M. Betrand II.
    (*C. R.* **122** (1896) 1314 = *WA* III.118)
Entgegnung auf die wärmetheoretischen Betrachtungen des Hrn. E. Zermelo
    (*Wied. Ann.* **57** (1896) 773–784 = *WA* III. 119)
Ein Wort der Mathematik an die Energetik
    (*Ann. d. Phys. u. chem.* **57** (1896) 39 = *PS* 8, pp. 104–136)
Zur Energetik
    (*Ann. d. Phys. u. chem.* **58** (1896) 595 = *PS* 9, pp. 137–140, translated in the present volume)
Röntgens neue Strahlen
    (*Elektro-Techniker* **14** (1896) 385 = *PS* 13, pp. 188–197)
1897 *Vorlesungen über die Principe der Mechanik, I. Theil*
    (J. A. Barth, Leipzig, 1897, some sections translated in the present volume)
Preface to Charles E. Curry *Theory of Electricity and Magnetism*
    (London and New York 1897, English original; Curry's work is based on Boltzmann's *Vorlesungen über die Maxwellsche Theorie* etc., 1891 und 1893)

Zu Hrn. Zermelos Abhandlung,, Über die mechanische Erklärung irreversibler Vorgänge".
(*Wied. Ann.* **60** (1897) 392–398 = *WA* III.**120**)
Über einen mechanischen Satz Poincaré's
(*Wien. Ber.* **106** (1897) 12–20 = *WA* III.**121**)
Über Rotationen im konstanten elektrischen Felde
(*Wied. Ann.* **60** (1897) 399–400 = *WA* III.**122**)
Über einige meiner weniger bekannten Abhandlungen über Gastheorie und deren Verhältnis zu derselben
(*Verh. der 69. Vers. D. Naturf. und Ärzte*, 19–26. Braunschweig 1897; *Jahresber. d. D. Mathem.-Vereinigung* **6**, I (1899) 130–138 = *WA* III.**123**)
Kleinigkeiten aus dem Gebiete der Mechanik
(*Verh. d. 69. Vers. D. Naturf. u. Ärzte* 26–29. Braunschweig 1897; *Jahresb. d. D. Mathem.-Vereinigung* **6**, I (1899) 138–142 = *WA* III.**124**)
Über irreversible Strahlungsvorgänge I.
(*Berl. Ber.* (1897) 660–662 = *WA* III.**125**)
Über irreversible Strahlungsvorgänge II.
(*Berl. Ber.* (1897) 1016–1018 = *WA* III.**126**)
Über die unentbehrlichkeit der Atomistik in der Naturwissenschaft
(*Ann. d. Phys. u. Chem.* **60** (1897) 231 = *PS* **10**, pp. 141–157, **translated in the present volume**)
Nochmals über die Atomistik
(*Ann. d. Phys. u. Chem.* **61** (1897) 790 = *PS* **11**, pp. 158–161, **translated in the present volume**)
Über die Frage nach der objektiven Existenz der Vorgänge in der unbelebten Natur
(*Wien. Ber.* **106** (1897) 83 = *PS* **12**, pp. 162–187, **translated in the present volume**)
Some Errata in Maxwell's paper 'On Faraday's Lines of Force'
(*Nature* **57** (1897) 77–79, **English original**)
1898 *Vorlesungen über Gastheorie II. Theil*
(J. A. Barth, Leipzig, 1898, **translated** by S. G. Brush as *Lectures on Gas Theory*, Berkeley, 1964)
ed.: J. C. Maxwell: *Über physikalische Kraftlinien*
(in Ostwalds Klassiker der exakten Wissenschaften, Leipzig)
Über vermeintlich irreversible Strahlungsvorgänge
(*Berl. Ber.* **127** (1898) 182–187 = *WA* III.**127**)
Über die sogenannte *H*-Kurve
(*Math. Ann.* **50** (1898) 325–332 = *WA* III.**128**)
Vorträge, gehalten beider 70. Versammlung Deutscher Naturforscher und Ärzte in Düsseldorf
(a) Zur Energetik
(b) Vorschlag zur Festlegung gewisser physikal. Ausdrücke
(c) Über die kinetische Ableitung der Formeln für den Druck des gesättigten Dampfes, für den Dissoziationsgrad von Gasen und für die Entropie eines das van der Waalsche Gesetz befolgenden Gases
(*Verh. d. 70. Vers. D. Naturf. u. Ärzte* (Düsseldorf 1898) 65–68, 74 = *WA* III.**129**)
1899 Sur le rapport des deux chaleurs spécifiques des gaz

(*C. R.* **127** (1898) 1009–1014 = *WA* III. **130**)
Über eine Modifikation der van der Waalschen Zustandsgleichung; gemeinschaftlich mit H. Mache
(*Wien. Anz.* **36** (16 March, 1899) 87–88; *Wied. Ann.* **68** (1899) 350–351 = *WA* III.**131**)
Über die Bedeutung der Konstante *b* des van der Waalschen Gesetzes; gemeinschaftlich mit H. Mache
(*Cambridge Phil. Trans.* **18** (1899) 91–93 = *WA* III.**132**)
Über die Zustandsgleichung van der Waals
(*Amsterdam. Ber.* (1899) 477-484 = *WA* III.**133**)
Über die Entwicklung der Methoden der theoretischen Physik in neuerer Zeit
(Lecture delivered to the Naturforscherversammlung, Munich, 22 September 1899 = *PS* **14**, pp. 198–277, **translated in the present volume**)
Über die Grundprinzipien und Grundgleichungen der Mechanik
(*Clark University 1889–1899, decennial celebration* (Worcester, Mass., 1899) pp. 261–309 = *PS* **16**, pp. 253–307, **first two lectures translated in the present volume**)

1900 Die Druckkräfte in der Hydrodynamik und die Hertzsche Mechanik
(*Ann. d. Phys.* (4) **1** (1900) 673–677 = *WA* III.**134**)
Zur Geschichte unserer Kenntnis der inneren Reibung und Wärmeleitung in verdünnten Gasen
(*Phys. Zeitschr.* **1** (1900) 213 = *WA* III.**135**)
Notiz über die Formel für den Druck der Gase
(Livre Jubilaire dedié à H. A. Lorentz (1900) 76–77 = *WA* III.**136**)
Eugen von Lommel
(*Jahresber. d. D. Mathem.-Vereinigung* **8** (1900) 47–53 = *WA* III.**137**)
Über die Prinzipien der Mechanik I
(Inaugural lecture at Leipzig, November 1900 (Leipzig, 1903) = *PS* **17**, pp. 309–330, **translated in the present volume**)

1902 Über die Form der Lagrangeschen Gleichungen für nichtholonome, generalisierte Koordinaten
(*Wien. Ber.* **111** (1902) 1603–1641 = *WA* III.**138**)
Über die Prinzipien der Mechanik II
(Inaugural lecture at Vienna, October 1902 (Leipzig, 1903) = *PS* **17**, pp. 330–337, **translated in the present volume**)
Models
(Encyclopaedia Britannica, 10th ed., 1902, s.v., **English original reprinted in the present volume** from the 11th ed., 1911, s.v. *Model*)

1903 Ein Antrittsvortrag über Naturphilosophie
(Inaugural lecture delivered on 26 October 1903, *Die Zeit* 11 December 1903 = *PS* **18**, pp. 338–344, **translated in the present volume**)
Besprechung des Lehrbuches der theoretischen Chemie von Wilhelm Vaubel
(Berlin 1903) (*PS* **21**, pp. 379–384)

1904 *Vorlesungen über die Prinzipe der Mechanik, II. Theil*
(J. A. Barth, Leipzig, 1904, **some sections translated in the present volume**)
Über das Exnersche Elektroskop. Gemeinsam mit A. Boltzmann
(*Wien, Anz.* **41** (3 November 1904) 325; *Phys. Ztschr.* **6** (1905) 2 = *WA* III. **139**)
Über statistische Mechanik

(*PS* 19, pp. 345–363, translated (by S. Epsteen) in *The Congress of Arts and Sciences, St. Louis* 1904 (Boston, 1905) under the title 'The Relations of Applied Mathematics' (scil. 'to physics') also translated in the present volume)
Entgegnung auf einen von Prof. Ostwald über das Glück gehaltenen Vortrag
(*PS* 20, pp. 364–384, translated in the present volume)

1905 *Populäre Schriften*
(J. A. Barth, Leipzig, 1905: contents indicated in this bibliography and in the contents-pages and editor's note to the present volume)
Über eine These Schopenhauers
(Lecture delivered to the philosophische Gesellschaft in Vienna, 21 January 1905, *PS* 22, pp. 385–402, translated in the present volume)
Reise eines deutschen Professors ins Eldorado
(*PS* 23, pp. 403–435)
Kinetische Theorie der Materie (with J. Nabl)
(In *Enzyklopädie der mathematischen Wissenschaften*, Leipzig, V-1. Heft 4, pp. 493–557)

1908 H. Buchholz: *Das mechanische Potential nach Vorlesungen von L. Boltzmann bearbeitet, und die Theorie der Figure der Erde*
(Leipzig, J. A. Barth)

1909 *Wissenschaftliche Abhandlungen*, 3 Volumes
(edited by F. Hasenöhrl (Leipzig, J. A. Barth): Contents indicated in this bibliography)

1916 H. Buchholz: *Angewandte Mathematik, mit einem Anhang über das elastische und das hydrodynamische Potential auf Grund von Vorlesungen L. Boltzmanns*
(Leipzig, J. A. Barth)

1920 *Vorlesungen über die Prinzipe der Mechanik, III. Theil*
(Elastizitätstheorie und Hydrodynamik, edited by H. Buchholz (separate edition of the appendix to Buchholz's 1916 volume), Leizpig, J. A. Barth)

1964 *Lectures on Gas Theory*
(translated by S. G. Brush, Berkeley, 1964; with an introduction and bibliography)

1974 *Theoretical Physics and Philosophical Problems*, selected writings (translated by P. Foulkes: the present volume)

II. A SELECTION OF WRITINGS DEDICATED TO L. BOLTZMANN OR DISCUSSING HIS LIFE OR WORK

1904 (ed.) S. Meyer: *Festschrift Ludwig Boltzmann gewidmet zum sechzigsten Geburtstag 20. Februar 1904*
(J. A. Barth, Leipzig) with a portrait; contains contributions by P. Duhem, M. Planck, E. Mach, J. Larmor, G. Frege, S. Arrhenius, and W. Nernst, among others.

1906 G. H. Bryan: Obituary of L. Boltzmann
(*Nature* 74 (1906) 569)

1909 P. Ehrenfest: Ludwig Boltzmann †
(*Math.-Nw. Blätter* 6 (1909) = Collected Scientific Papers (1959) 12)

1911 P. and T. Ehrenfest: *Begriffliche Grundlagen der statistischen Auffassung in der Mechanik*

(*Enzyklopädie der Math. Wiss.* IV.**32** (Leipzig 1911), translated by M. J.
Moravcsik as *The Conceptual Foundations of the Statistical Approach in
Mechanics* (Ithaca, N.Y., 1959))

1925 G. Jaeger: Ludwig Boltzmann
      (*Neue österreichische Biographie II* (1925) 117–137)

1944 E. Schrödinger: The Statistical Law in Nature
      (*Nature* **153** (1944) 704–5)
      A. Sommerfeld: *Das Werk Boltzmanns*
      L. Flamm: *Die Persönlichkeit Boltzmanns*
      (*Wiener Chem.-Zeitung* **47** (1944) 25 and 28)

1955 E. Broda: *Ludwig Boltzmann, Mensch, Physiker, Philosoph*
      (Vienna 1955, Berlin 1957, with a portrait)
      A. Sommerfeld: Boltzmann, Ludwig
      (*Neue deutsche Biographie* **2** (1955) 436–7)

1956 B. I. Davydov: *Lektsii po teorii gazov*
      (Moscow; translation into Russian of *Vorlesungen über Gastheorie* (1896–8)
      with introduction and notes)

1957 L. Flamm: In Memory of Ludwig Boltzmann
      N. N. Bogolyubov and Yu. V. Sanochkin: Ludwig Boltzmann
      B. I. Davydov: A Great Physicist
      (*Uspekhi Fizicheskikh Nauk* **61** (1957) 3, 7, and 17 = *Advances in Physical
      Sciences* **61** (1957) 3, 7, and 20)
      H. Thirring: Ludwig Boltzmann in seiner Zeit
      (*Naturwiss. Rundsch.* **10** (1957) 411–5

1959 R. Dugas: *La théorie physique au sens de Boltzmann et ses prolongements
      modernes*
      (Neuchâtel, 1959; with an introduction by L. de Broglie)

1964 S. G. Brush: *Lectures on Gas Theory*
      (Berkeley; translation of *Vorlesungen über Gastheorie* (1896–8) with in-
      troduction and bibliography)

1967 S. G. Brush: Foundations of Statistical Mechanics
      (*Arch. for Hist. of Exact Sci.* **4** (1967) 145–183)

1970 M. Tomáš: *Filosoficky portrét Ludwiga Boltzmanna*
      (Prague, 1970)
      S. G. Brush: Boltzmann, Ludwig
      (*Dictionary of Scientific Biography* II (1970) 260–8)

1971 E. Hiebert: The Energetics Controversy and the New Thermodynamics
      (in *Perspectives in the History of Science and Technology* (ed. D. H. D.
      Roller), Norman, Oklahoma, 67–86)

1973 D. Flamm: Life and Personality of Ludwig Boltzmann
      E. Broda: Philosophical Biography of Ludwig Boltzmann
      M. J. Klein: The Development of Boltzmann's Statistical Ideas
      (*The Boltzmann Equation – Acta Physica Austriaca, Supplementum* **10**
      (1973) 3, 17, and 53)

INDEX OF NAMES

VIENNA CIRCLE COLLECTION